MORE AD
EQUIPMENT

Other Titles of Interest

BP70	Transistor Radio Fault Finding Chart
BP101	How to Identify Unmarked IC's
BP110	How to Get Your Electronic Projects Working
BP120	Audio Amplifier Fault Finding Chart
BP239	Getting the Most From Your Multimeter
BP248	Test Equipment Construction
BP265	More Advanced Uses of the Multimeter
BP267	How to Use Oscilloscopes and Other Test Equipment

MORE ADVANCED TEST EQUIPMENT CONSTRUCTION

by

R. A. PENFOLD

BERNARD BABANI (publishing) LTD
THE GRAMPIANS
SHEPHERDS BUSH ROAD
LONDON W6 7NF
ENGLAND

Please Note

Although every care has been taken with the production of this book to ensure that any projects, designs, modifications and/or programs etc. contained herewith, operate in a correct and safe manner and also that any components specified are normally available in Great Britain, the Publishers do not accept responsibility in any way for the failure, including fault in design, of any project, design, modification or program to work correctly or to cause damage to any other equipment that it may be connected to or used in conjunction with, or in respect of any other damage or injury that may be so caused, nor do the Publishers accept responsibility in any way for the failure to obtain specified components.

Notice is also given that if equipment that is still under warranty is modified in any way or used or connected with home-built equipment then that warranty may be void.

First Published – November 1989

British Library Cataloguing in Publication Data

Penfold, R. A.

More advanced test equipment construction

1. Electronic testing equipment. Construction

I. Title

621.3815'48

ISBN 0 85934 194 1

Printed and bound in Great Britain by Cox & Wyman Ltd, Reading

Preface

Anyone involved with electronic project construction soon realises the importance of having a reasonable range of test equipment. Newly constructed projects can and do fail to work from time to time, and operational projects can develop faults. Without test equipment it is possible to track down wiring errors, accidental short circuits, etc., but difficult faults such as "dud" components can be very difficult and time consuming to track down without the aid of some test gear.

The book "Test Equipment Construction" (BP248) describes some simple but useful items of test equipment that are designed to fill in most of the gaps in the average multimeter's repertoire. This publication provides further designs for the electronics hobbyists' workshop. In some cases these are more sophisticated versions of equipment featured in the earlier book, but in most cases they are for equipment of a type not featured in it. A range of useful digital measuring instruments are described in Chapter 1, with a few miscellaneous designs being provided in Chapter 2. The projects include a current tracer, a high quality digital capacitance meter, a dynamic transistor tester, and a crystal calibrator.

Unlike "Test Equipment Construction" (BP248), this book is not primarily aimed at beginners and near beginners at electronic project construction. A certain amount of practical experience and knowledge of electronics theory is assumed. However, you do not need to be an expert at electronics in order to build and use these designs, and anyone with a moderate amount of electronics experience should be able to tackle these projects. Someone who has gained some experience by building and using projects from BP248 should be able to successfully undertake and utilize the designs featured here.

R. A. Penfold

Warning

Never make tests on any mains powered equipment that is plugged into the mains unless you are quite sure you know exactly what you are doing.

Remember that capacitors can hold their charge for some considerable time even when equipment has been switched off and unplugged.

Contents

Chapter 1

DIGITAL MEASURING EQUIPMENT

Test equipment falls into two main categories – those devices that measure something and those that generate signals of one kind or another. In this first chapter we will consider several pieces of measuring equipment. Something they all have in common is that they are digital instruments. In fact they are all based on the same three and a half digit d.v.m. (digital voltmeter) module. First the digital voltmeter module will be described, and then we will move on to a range of add-ons that will enable it to perform a range of useful functions.

Half Digits

There are several digital voltmeter integrated circuits available, and these enable highly accurate voltmeters to be produced using a minimal number of components. It is not actually all that difficult to produce a d.v.m. from logic circuits and linear devices, but a special d.v.m. integrated circuit will generally provide better accuracy at lower cost, and with a much lower parts count. The less expensive d.v.m. chips are designed to drive three and a half digit displays. In other words, a four digit display where the most significant digit can only be zero or one. In fact it is more accurate to say that the most significant digit is either blanked out or is set at 1. It has only two segments, and can not display a leading zero (but presumably would never need to anyway).

The use of a so-called half digit might seem a bit gimmicky, but it is really quite a good arrangement. With digital test equipment it is normal to have each measuring range one decade higher than the previous one. A digital resistance meter for instance, might have ranges of 999 ohms, 9.99k, 99.9k, 999k, and 9.99M full scale. A slight problem with this arrangement is that some values produce an over-range error on cne range, but quite a low reading on the next range up. For instance, a resistor having a value of 1024 ohms would give an over-range indication on the 999 ohms range, and a far from full scale reading of 1.02k on the 9.99k range. This tends to give limited accuracy on these awkward values, although it has to be said that for most purposes results would still be perfectly adequate. Anyway, the extra half digit enables these values to be brought in-range on the lower range, giving better accuracy.

Four and a half digit d.v.m. chips offer extremely good specifications, and give the potential for very high degrees of accuracy. They are relatively expensive though, and in practical applications it is very difficult to produce an overall design that justifies the extra digit. With a three and a half digit display each increment of the least significant digit represents just 0.05% of the full scale value. The equivalent figure for a four and a half digit display is 0.005%! With the circuits preceding the d.v.m. being built using what will at best be 1% tolerance components, even a three and a half digit display could be regarded as being over-specified. There is no realistic prospect of producing circuits to exploit the full resolution and accuracy of a four and a half digit d.v.m. The circuits in this book are therefore based on an ordinary three and a half digit d.v.m. design.

Chips to drive light emitting diode (l.e.d.) and liquid crystal displays (l.c.d.) are available. As far as using the instruments is concerned, l.e.d.s are better in dim lighting while l.c.d.s are superior under bright conditions. There is little overall advantage to either type in this respect. I chose to base these circuits on a chip which operates with a liquid crystal display simply because this type of display requires a much lower operating current. In fact the whole d.v.m. circuit draws a supply current of only about two milliamps or so. This permits economic battery operation, and even a small (PP3 size) 9 volt battery will have a long operating life. The current consumption for the l.e.d. equivalent of this d.v.m. circuit would be largely dependent on the number of segments switched on at the time, and the particular l.e.d. segment current used, but would probably exceed one hundred milliamps on occasions. This would necessitate the use of a high capacity battery which would still have a relatively limited operating life.

Single Integration

Digital voltmeters almost invariably use the single or dual

slope integration techniques. The ICL7106 d.v.m. chip used in the designs described in this chapter is of the single slope integration variety. This is somewhat inferior to the dual slope type in certain respects, but it is nevertheless capable of providing excellent results. The single slope integration process operates by having a clock oscillator driving the input of a counter circuit via a gate circuit. In practical systems the counter drives the display via latches which have their contents up-dated each time a new value is available from the counter. This gives a continuous display of the input voltage, with the counting process not being apparent to the user.

In order to give a reading of the input voltage, some means of converting the input voltage to a gate pulse of proportional length is required. This is achieved using an integrator, which is a circuit that charges a capacitor at a rate that is proportional to the input voltage. There is more than one way of using an integrator in a d.v.m., one of which is to compare the output voltage of the integrator with the input voltage. An input pulse to the integrator starts its output rising in voltage at a linear rate. When the output voltage of the integrator reaches the same potential as the input voltage, the input pulse is ended. The input pulse can be generated using a flip/flop which is set by the control logic circuit to start the input pulse, and reset by the comparator to end it.

The point to note here is that the higher the input voltage, the longer it will take the output voltage of the integrator to reach the input potential. As the integrator's output rises at a linear rate, the duration of the integrator's input pulse is proportional to the input voltage, and can therefore be used as the gate pulse. The block diagram of Figure 1.1 helps to explain the way in which the various stages interconnect, but this is in somewhat over-simplified form. In practice some fairly complex control logic circuits are needed in order to get everything functioning properly.

D.V.M. Circuit

The full circuit diagram of the digital voltmeter module appears in Figure 1.2. R5 and C3 merely form a lowpass filter at the input of the circuit. It is standard practice to include a filter of this type at the input of d.v.m.s since any noise on

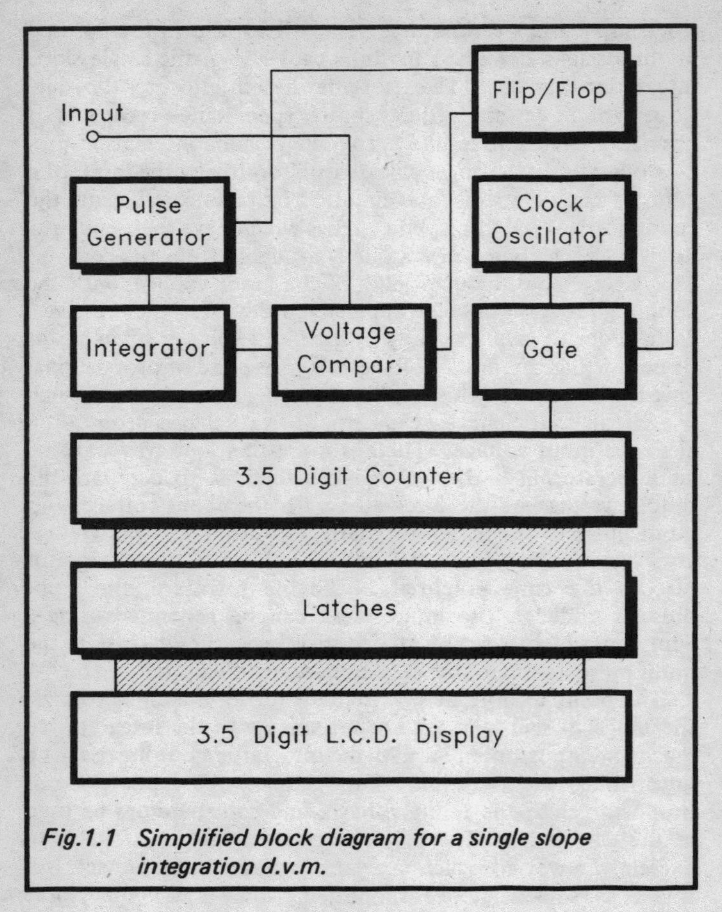

Fig.1.1 Simplified block diagram for a single slope integration d.v.m.

the input tends to cause variations in the displayed reading. This can make the display very difficult to read, and a lowpass filter with a low cut-off frequency greatly eases the problem by smoothing out the variations in the input voltage so that much more stable readings are obtained. A fairly low cut-off frequency is desirable as it gives a large amount of smoothing and very stable readings. On the other hand, this results in a significant delay between a new input voltage being applied to

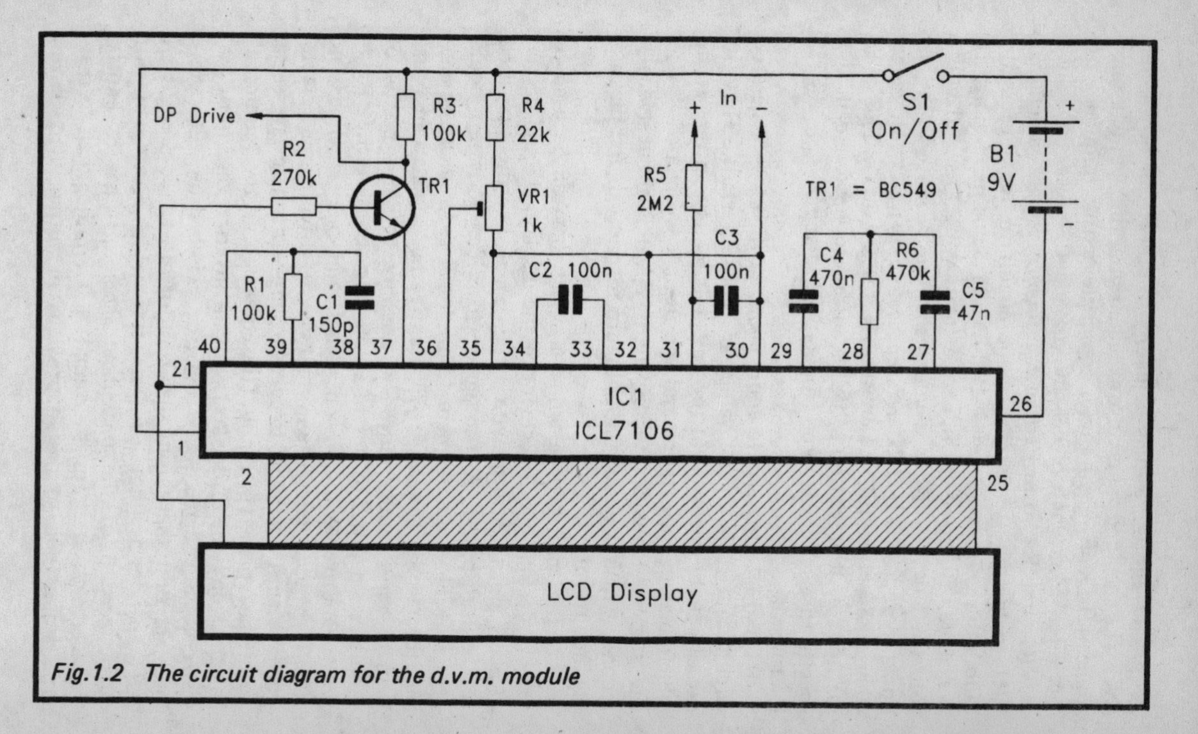

Fig.1.2 The circuit diagram for the d.v.m. module

the circuit and the display adjusting to reflect this new input level. The specified values give excellent smoothing, but in some applications you might prefer to set the cutoff frequency of the filter somewhat higher by making C3 lower in value.

VR1 and R4 form part of a reference voltage generator. VR1 enables the sensitivity of the circuit to be adjusted for calibration purposes, and a fairly wide adjustment range is available. The nominal full scale input voltage is 200 millivolts (or 199.9 millivolts if you wish to be pedantic). In this book there will often be references to (say) a 2 volt range when the actual full scale value is 1.999 volts. This is a method of abbreviation which is often to be found in the specifications sheets and operating manuals of digital instruments.

R1 and C1 are discrete components in the clock generator circuit, and their precise values are not important. They do not affect the accuracy of the unit, but merely set the rate at which readings are taken and the display is updated. With the suggested values the display is updated about twice a second, which is about the optimum for most digital measuring instruments.

R2, R3, and TR1 generate a signal that can be used to drive a decimal point segment of the display. Liquid crystal displays are not driven with a d.c. signal, like the ones used for l.e.d. displays. A d.c. signal will in fact activate a liquid crystal display, but segment "burn" will soon occur, rendering the display useless. Apparently d.c. signals can cause damage very rapidly indeed – possibly within a matter of minutes. In order to obtain a long operating life the display must be driven with an a.c. signal, and one which has no significant d.c. bias. The normal way of driving liquid crystal displays with a suitable signal is to apply a squarewave signal to the backplane (BP) input (the equivalent of the "common" terminal on a l.e.d. display), and an inverted version of that signal to any segment that must be switched on. Electronic switches turn the output signals on or off, as required. Here TR1 operates as a simple inverter stage which provides an antiphase version of the backplane signal. It has its emitter connected to the "Test" terminal of IC1 rather than to the 0 volt supply rail, so that it provides an output signal having the correct voltage range.

Most of the other components are either part of the integrator or the automatic zeroing circuit. Due to the inclusion of an effective automatic zeroing circuit, no manual zero adjustments at all are required (but there may be voltage offsets in the circuit driving the d.v.m. which will need to be manually nulled).

S1 is the on/off switch and B1 is the 9 volt battery supply. Note that some of the add-on circuits for the d.v.m. require a stabilised supply, and (or) a slightly higher supply voltage. For these the circuit of Figure 1.3 must be used. This is

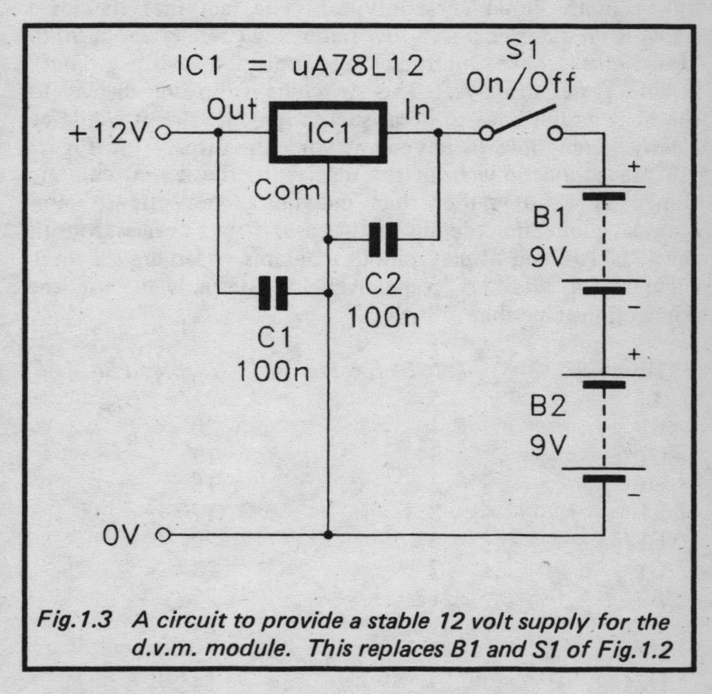

Fig.1.3 A circuit to provide a stable 12 volt supply for the d.v.m. module. This replaces B1 and S1 of Fig.1.2

simply a small 12 volt monolithic voltage regulator driven by two 9 volt batteries wired in series. This gives a well stabilised 12 volt output from the nominal 18 volts provided by the batteries. Note that the absolute maximum supply voltage rating of the ICL7106 d.v.m. chip is 15 volts, and it is not safe to use the 18 volt battery supply without the voltage regulator. It is a good idea to check that the correct 12 volt supply is being obtained from the finished circuit prior to fitting the d.v.m. chip into its socket.

Interconnections

The d.v.m. chip connects to the display via more than twenty wires. This large number of connections is necessary because the display can not be multiplexed. Although multiplexing is quite common with large l.e.d. displays, it is virtually unknown with liquid crystal types. The fact that the drive signal is an a.c. signal at a low frequency (usually about 50 to 100 hertz) makes multiplexing impossible with ordinary liquid crystal displays. The switching from one display to the next would have to be at such a low rate that it would be clearly perceptible to anyone viewing the display. In Figure 1.2 the connections from the display to the d.v.m. chip are shown as a bus rather than individual connections. For concise connection details between the two devices consult the ICL7106 and display pinout diagrams which are shown in Figures 1.4 and 1.5 respectively. Alternatively, use the connection table that follows.

Terminal	*ICL7106 Pin*	*Display Pin*
A1	5	21
B1	4	20
C1	3	19
D1	2	18
E1	8	17
F1	6	22
G1	7	23
A2	12	25
B2	11	24
C2	10	15
D2	9	14
E2	14	13
F2	13	26
G2	25	27

continued

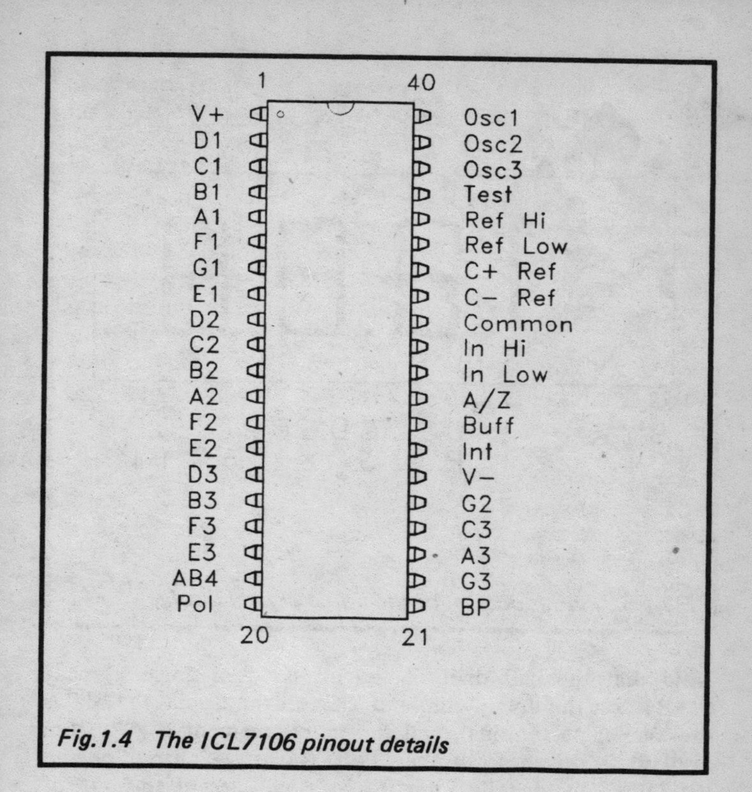

Fig.1.4 The ICL7106 pinout details

Terminal	*ICL7106 Pin*	*Display Pin*
A3	23	30
B3	16	29
C3	24	11
D3	15	10
E3	18	9
F3	17	31
G3	22	32
AB4/K	19	3
BP	21	1
Pol/Y	20	2

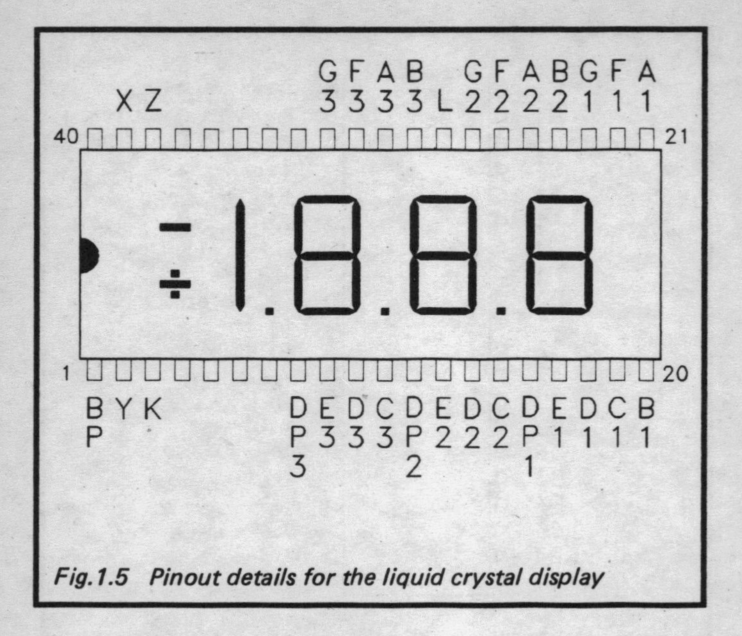

Fig.1.5 Pinout details for the liquid crystal display

Note that the half digit of the display is a single segment ("AB4" on the d.v.m. chip and "K" on the display). "BP" is the backplane terminal, and the minus segment ("Y" of the display) is driven from the "Pol" (polarity) output of the ICL7106. Incidentally, the ICL7106 will accept an input of either polarity, with the "–" segment being automatically activated if there is a reverse polarity input. This is very useful when the unit is used as a voltmeter, but may be inappropriate when it is used in certain types of test equipment. In the case of a transistor tester for example, the gain can never be a negative figure. Accordingly, the "Y" segment of the display should be left unconnected when the d.v.m. module is used in a unit of this type.

Most three and a half digit liquid crystal displays have other segments, such as "+" and "Bat" types, but these are left unused in this case. It should be pointed out that the pinout details of Figure 1.5 are correct for the display I used, and for the three and a half digit liquid crystal displays offered by the main component retailers. However, there could be other types which are suitable for use in this d.v.m. module but which have a different pinout configuration. If possible, check the pinout diagram for the display you intend to use, and if necessary modify the connections to suit the different pinout arrangement.

When constructing a unit based on the d.v.m. module remember that the ICL7106 is a MOS device and therefore requires the standard anti-static handling precautions to be observed. Although a lot of constructors often ignore these precautions (myself included), as the ICL7106 is not a particularly cheap component, it would be as well to treat it with due respect. It should certainly be fitted in a holder, and should not be fitted into place until the project is in all other respects finished. Until then it should be left in its anti-static packaging (conductive foam, plastic tube, etc.).

The liquid crystal display is also a relatively delicate component that needs to be treated with respect. Apart from being easily damaged electrically, they are often physically something less than strong. I would strongly urge that the display should be fitted in a holder, but obtaining a suitable type could prove difficult. Although the display has a 40 pin d.i.l. encapsulation, it is not of the standard 40 pin type with 0.6 inch pin spacing. The two rows of pins are spaced some 1.4 inches apart. Probably the best solution to the problem is to cut an ordinary 40 pin d.i.l. integrated circuit holder into two 20 pin s.i.l. strips which can then be mounted on the circuit board with the appropriate spacing. If you can obtain "Soldercon" pins these should be suitable, since they can be cut into rows having any desired number of pins, and mounted with any required spacing.

The display should be mounted behind a suitable cutout in the front panel of the unit, and it is advisable to fit some clear plastic material behind the cutout to provide some protection for the front of the display. In Figure 1.5 the display is shown as having its orientation marked via the usual half-round at the pin 1 end of the component. On many displays the actual marking is just a black line, and in a few cases there seems to be no marking at all. This does not really matter, since a close visual inspection of the display will reveal the segment pattern, and this enables the correct orientation for the device to be determined.

In the components list a multi-turn preset ("trimpot") has been specified for VR1. An ordinary miniature preset resistor is usable, but it will be very difficult to accurately calibrate the unit using one of these. It is just possible that an ordinary preset would not allow precisely the required value to be set. A multi-turn type is much better, and will enable the correct reading to be easily set.

One final point is that the d.v.m. chip incorporates overload indication. If the input voltage is too high, the half digit is switched on, and the three full digits are turned off. The polarity indicator still operates (i.e. "1" is displayed for a positive overload: "−1" is displayed for a negative overload).

Components for D.V.M. Module (Fig.1.2)

Resistors (all 0.25 watt 5% or better)

R1	100k
R2	270k
R3	100k
R4	22k
R5	2M2
R6	470k

Potentiometer

VR1	1k multi-turn preset

Capacitors

C1	150p ceramic plate
C2	100n polyester
C3	100n polyester
C4	470n polyester
C5	47n polyester

Semiconductors

IC1	ICL7106
TR1	BC549
Display	3½ digit liquid crystal type

Miscellaneous

B1	9 volt (PP3 size)
S1	s.p.s.t. miniature toggle
	Battery connector
	40 pin d.i.l. holder
	Holder for display (see text)
	Clear plastic for "window"

Components for 12 Volt Stabilized Supply (Fig.1.3)

Capacitors

C1	100n ceramic
C2	100n ceramic

Semiconductors

IC1	μA78L12 (12 volt 100mA positive voltage regulator)

Miscellaneous

B1	9 volt (PP3 size)
B2	9 volt (PP3 size)
S1	s.p.s.t. miniature toggle

(Note that S1 and B1 are not additional components, since they effectively replace S1 and B1 in the main d.v.m. circuit.)

High Resistance Voltmeter

The obvious test gear application for a d.v.m. is in a high resistance voltmeter. If you already have a digital multimeter there is not much point in building a digital high resistance voltmeter, since you effectively have one of these already when you use the multimeter on a d.c. voltage range. If you have an analogue multimeter, then this probably has an input resistance of 20k per full scale volt, and a relatively low input resistance on all but the highest voltage ranges. A high resistance voltmeter will then be a very useful addition to your test gear, as it will provide more reliable measurements. Apart from the better accuracy provided by a digital instrument anyway, it has the advantage of giving reduced loading on the test point.

This loading occurs because the voltmeter taps off some of the current in the test circuit. The higher the resistance through the voltmeter, the less current that is tapped off, and the lower the loading effect. A standard 20k/volt analogue multimeter is based on a meter movement having a full scale value of 50 microamps. The current tapped off from the test circuit is therefore quite low, and will not exceed 50 microamps. On the other hand, the current flow in parts of some circuits is extremely low indeed. In the base circuit of a transistor for instance, the current flow is often only around 1 to 5 microamps. Even though the voltage present at the test point might be something approaching the full scale value of the voltmeter, a very low reading will be obtained because there is insufficient current at the test point to drive the meter properly.

The circuit diagram of Figure 1.6 helps to explain exactly what happens. Here we have a simple emitter follower stage with base biasing provided by R1 and R2. These bias the

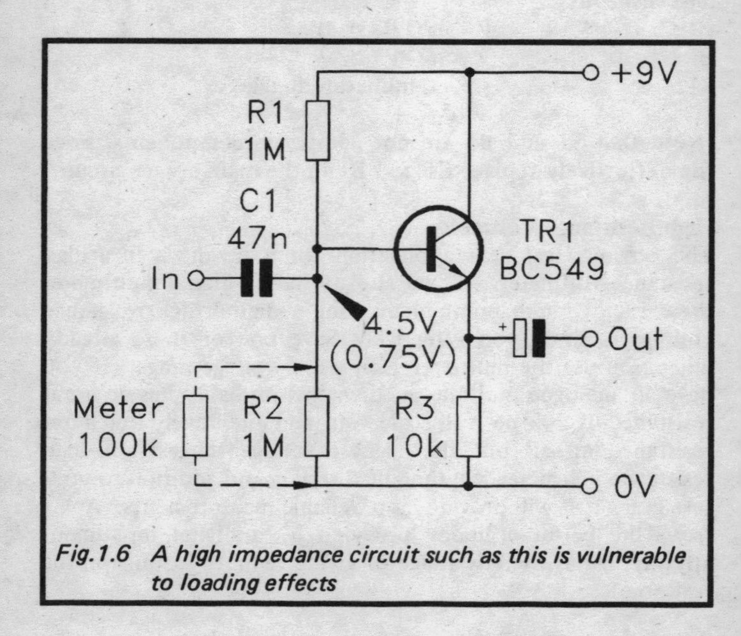

Fig.1.6 A high impedance circuit such as this is vulnerable to loading effects

input of the amplifier to about half the supply voltage. In theory the voltage they provide is exactly half the supply potential, but in practice the tolerances of the two resistors must be taken into account, and the input resistance of TR1 shunts R2 and reduces the bias level slightly. For the sake of this example we will ignore the effect of TR1, and will assume that there is half the supply voltage (4.5 volts) at the base of TR1.

Connecting the multimeter from the 0 volt supply to the junction of R1 and R2 to measure this voltage upsets the biasing by effectively adding a resistor in parallel with R2. The value of this resistor depends on the sensitivity of the multimeter and the voltage range in use. Using a 20k/volt instrument on the 5 volt range to make the measurement would place 100k (5 volts x 20k/volt = 100k) in parallel with R2. The shunting effect of a 100k resistor on a 1M type is clearly going to be quite dramatic. The combined resistance of the two components must be less than the 100k of the meter, and if you work it out you should arrive at an answer of just under 91k. This would reduce the voltage at the test point to less than one-tenth of the supply voltage, giving a reading on the meter of well under 1 volt. Note that strictly speaking the meter is not giving an incorrect reading. It accurately registers the voltage present at the test point. This voltage is present only while the meter is connected to the test circuit though, and in this respect the reading is an erroneous one.

A high resistance voltmeter reduces the problem by using an active device at its input in order to reduce the input current drawn from the circuit under test. The input current drawn by the d.v.m. module is only about 1pA (i.e. one-millionth of a microamp), which is totally insignificant. A practical voltmeter circuit must have several measuring ranges though, and this necessitates the addition of an attenuator at the input of the circuit. The attenuator must draw much more current than the input current of the meter circuit, so that loading of the attenuator and consequent inaccuracies are avoided.

Most digital voltmeters have an input resistance of about 10 or 11 megohms. A substantially higher input resistance could probably be used without any risk of loading the attenuator, but there are problems in obtaining high stability close tolerance resistors having the very high values that would be needed. In practice an input resistance of about 10 to 11 megohms is sufficient to ensure good accuracy. Returning to our example circuit of Figure 1.6, loading this with a resistance of about 10 to 11 megohms will produce a significant drop in the voltage at the test point, but only by a few per-cent. This will give a reading of adequate accuracy to determine whether or not there is a fault at the test point.

Attenuator Circuit

An attenuator circuit to convert the d.v.m. module into a multi-range high resistance voltmeter is shown in Figure 1.7. This is a standard four step type providing attenuations of 0dB, 20dB, 40dB, and 60dB, or attenuation factors of 1, 10, 100, and 1000 if you prefer. This gives four measuring ranges having full scale values of 0.2 volts, 2 volts, 20 volts, and 200 volts. The input resistance is a little over 11 megohms. R6, D1 and D2 clip excessive input voltages at about plus and minus 0.7 volts, and protect the d.v.m. chip against damage by serious overloads.

If you wish to have the correct decimal point segment for each range automatically activated, a "spare" pole of S1 must be used to switch the DP drive signal through to the correct decimal point segments of the display. The circuit of Figure 1.8 shows a suitable method of decimal point switching. On the lowest range this displays the reading in millivolts and not volts. It is not possible to have the reading in volts on this range, because there is no decimal point segment to the left of the leading digit (or half digit to be precise).

In order to live up to the potential accuracy of the d.v.m. module it is important that the attenuator resistors are close tolerance types. The higher their accuracy the better, but in practice you are unlikely to be able to get components having tolerances of better than 1%. These will provide good results, but will slightly limit the accuracy of the unit on ranges other than the one on which the unit is calibrated.

Calibration is very simple, but you need an accurate voltage source against which the unit can be set up. The calibration voltage should represent about 50 to 100% of the full scale

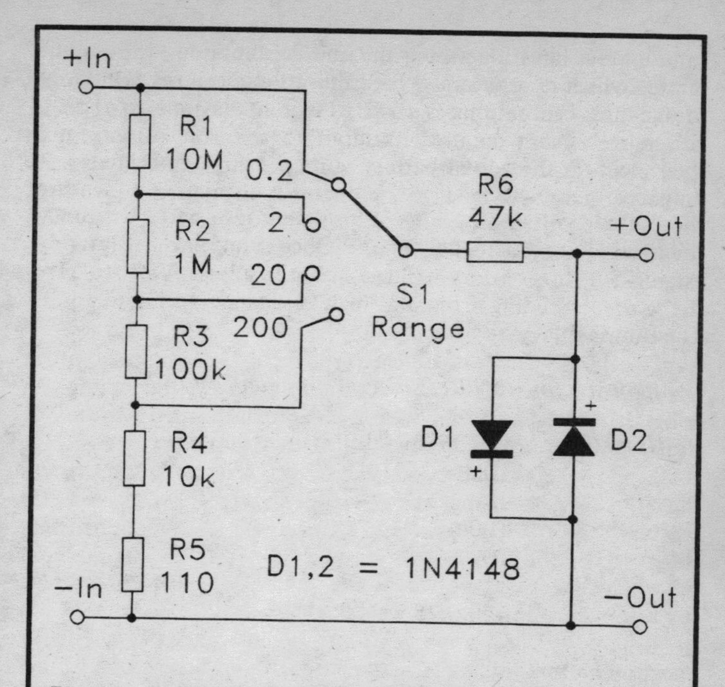

Fig.1.7 The input attenuator to convert the d.v.m. into a multi-range high resistance voltmeter

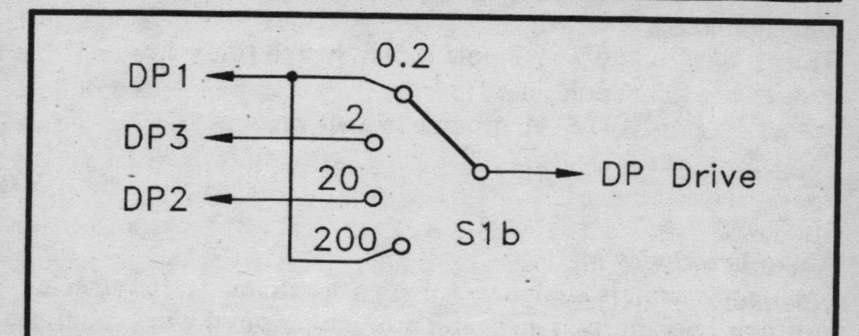

Fig.1.8 The decimal point switching for the high resistance voltmeter

value of the range on which the unit is calibrated. It does not matter which range you use for calibration purposes. One way of tackling calibrations is to first switch the unit to the 20 volt range. Next connect its input to a 9 volt battery, and then measure the actual battery voltage using a multimeter. If you have access to one, use a digital multimeter to measure the battery voltage, as this will probably give better accuracy than using an analogue type. Once you have accurately established the battery voltage, simply adjust VR1 to give the correct reading from the high resistance voltmeter unit. The unit is then ready for use.

Components for High Resistance Voltmeter (Fig. 1.7)

Resistors (all 0.6 watt 1% metal film unless noted)

R1	10M
R2	1M
R3	100k
R4	10k
R5	110R
R6	47k 0.25 watt 5%

Semiconductors

D1	1N4148
D2	1N4148

Miscellaneous

S1	4 way 3 pole rotary switch (only two poles used)
	D.V.M. module (9 volt version)

Audio Frequency Meter

Frequency meters designed for high accuracy at frequencies into the megahertz range are not based on d.v.m. circuits. They use a system of pulse counting with highly accurate gate periods controlled by a crystal clock oscillator. For audio frequency use though, quite good results can be obtained using a d.v.m. preceded by a frequency to voltage converter.

Even quite simple frequency to voltage converters can give good results at the relatively low frequencies involved in this application. The unit featured here has four frequency ranges, as follows:–

Range 1	–	0 – 199.9Hz
Range 2	–	0 – 1.999kHz
Range 3	–	0 – 19.99kHz
Range 4	–	0 – 199.9kHz

The arrangement used in the frequency to voltage converter featured here is shown in the block diagram of Figure 1.9. The buffer amplifier at the input is needed to give the unit a high input impedance so that it does not significantly load the signal source. It gives an input impedance of about 500k (somewhat less at high frequencies due to the inevitable input capacitance). The voltage amplifier boosts the sensitivity of the circuit to a more useful level. A minimum input level of about 25 millivolts r.m.s. is needed in order to drive the unit properly.

A trigger circuit is driven from the voltage amplifier, and this provides three functions. Firstly it provides an output signal having the fast rise and fall times needed to drive the next stage properly. Secondly, it ensures that slightly inadequate input levels do not cause misleading readings to be obtained. If the input level is slightly too low to drive the unit properly there will be no output from the trigger, and zero reading from the unit. Thirdly, it helps to avoid problems with noise on the input signal giving erroneous results.

The final stage of the unit is a monostable multivibrator, and it is this stage that provides the frequency to voltage conversion. The output of the monostable is a pulse having a duration that is set by a C – R timing circuit, and which is independent of the input pulse duration. The waveform diagram of Figure 1.10 helps to explain the way in which the frequency to voltage conversion is provided. Waveform (a) has a 1 to 8 mark space ratio, and the average voltage is therefore 12.5% of the V+ level. In (b) the input frequency has been doubled so that there are twice as many pulses in a

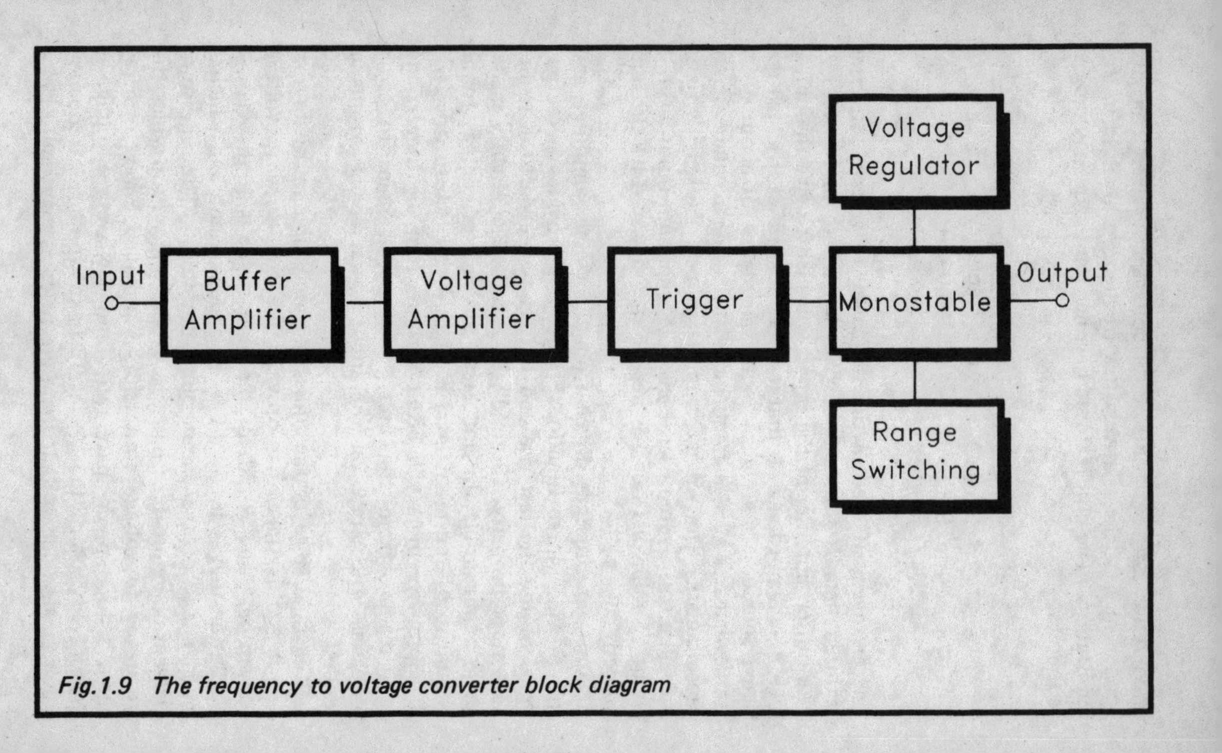

Fig.1.9 The frequency to voltage converter block diagram

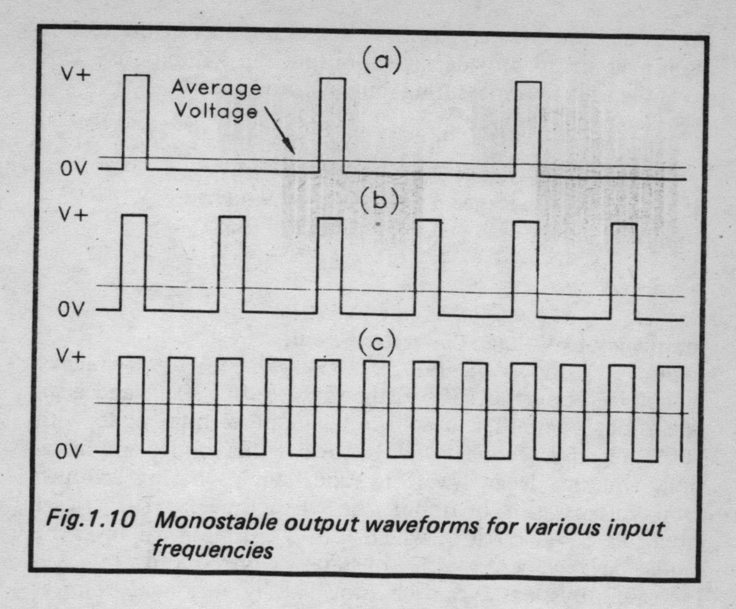

Fig.1.10 Monostable output waveforms for various input frequencies

given period of time, and the mark space ratio is 1 to 4. This gives an average voltage equal to 25% of the V+ level. In (c) the input frequency has once again been doubled, taking the mark space ratio to 1 to 1, and giving an average output voltage that is equal to 50% of the V+ level. The circuit thus provides the required action, with a linear relationship between the input frequency and the average output voltage.

Although strictly speaking the circuit is not a frequency to voltage converter, since the output is a series of pulses and not a d.c. level, all that is needed in order to give a true frequency to voltage conversion is a lowpass filter. This smooths out the pulses to give a low ripple d.c. output potential equal to the average output voltage. In this case there is no need to use a filter at the output of the monostable, since a suitable filter is included at the input of the d.v.m. module.

The monostable must be powered from a stable supply as any change in the supply voltage will be reflected in a proportional shift in its average output voltage. Several measuring ranges are provided by using switched resistors in the C – R timing circuit to provide several output pulse durations. This gives the unit four measuring ranges, as follows:–

Range 1	–	0 to 199.9Hz
Range 2	–	0 to 1.99kHz
Range 3	–	0 to 19.99kHz
Range 4	–	0 to 199.9kHz

Frequency to Voltage Converter Circuit

Figure 1.11 shows the full circuit diagram for the frequency to voltage converter. IC4 is the input buffer stage, and is an operational amplifier used in the non-inverting mode with 100% negative feedback. IC3 provides the voltage amplification, and this is an inverting mode circuit having an open loop voltage gain of about 26dB (20 times). The trigger circuit is based on IC2, which uses a conventional inverting mode arrangement with hysteresis introduced by R9. Increased hysteresis can be provided by making R9 lower in value. This gives better reliability on signals that contain a lot of noise, but at the cost of reduced sensitivity.

The monostable uses 555 timer IC1 in the conventional monostable mode. Actually IC1 is a low power version of the 555 timer, the TLC555CP. The output of the interface must be referenced to the negative input terminal of the voltmeter, not the 0 volt supply rail (which is several volts negative of the negative input). This often means that any circuit added ahead of the d.v.m. module must be powered from between the negative input terminal and the positive supply rail. Alternatively, and as in this case, the main circuit can be supplied from the normal supply rails, with only the output stage using the negative input as its 0 volt supply rail. This gives a relatively low supply voltage for the circuit or output stage, and there is the added problem that it must not draw a very high supply current if it is to leave the d.v.m. module functioning accurately. The TLC555CP is therefore a better choice for this application than the standard 555 timer.

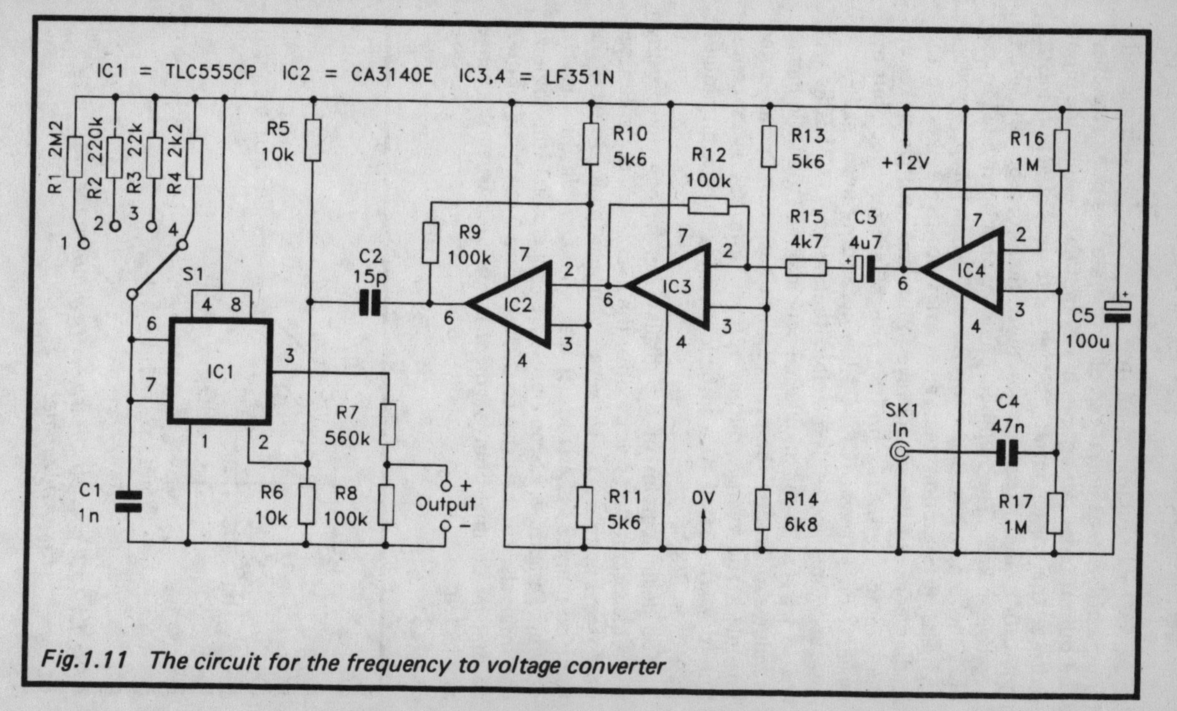

Fig.1.11 The circuit for the frequency to voltage converter

A 555 monostable is of the negative edge retriggerable type, which means that it can only operate as a pulse stretcher. The output pulse can not be shorter in duration than the input pulse at the trigger input. To give the desired action in this application the trigger pulses must therefore be very brief negative spikes. The required pulse shaping is provided by C2 plus the input impedance of R5 and R6. These provide a highpass filter action that gives brief positive spikes on the leading edges of the squarewave input, and negative spikes on the trailing edges. The positive pulses have no effect, but the negative ones trigger IC1. C2 has necessarily been made very low in value, and it is possible that some TLC555CPs will fail to trigger. If this should happen, making C2 a little higher in value should cure the problem, but keep its value as low as possible. The average output voltage from IC1 is excessive, but it is attenuated to a suitable level for the d.v.m. module by R7 and R8.

The decimal point of the display can be controlled using the same circuit that was used for the high resistance voltmeter unit (see Figure 1.8). However, you may prefer to have the units on range 2 as hertz rather than kilohertz. In this case the circuit of Figure 1.12 can be used. This is much the same as the original, but the drive to DP3 on range 2 has been omitted so that no decimal point segment is activated when this range is selected.

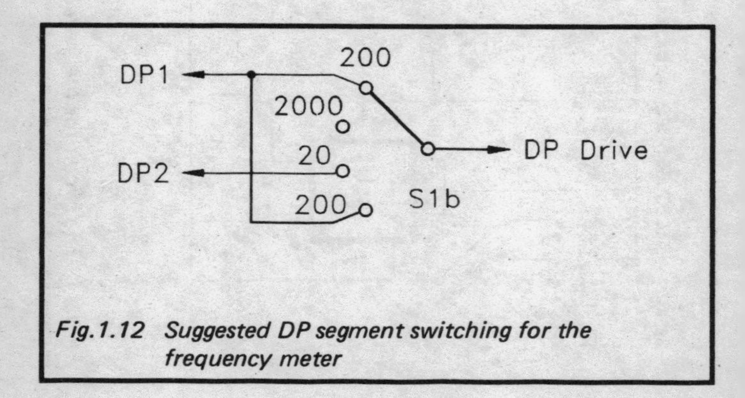

Fig.1.12 Suggested DP segment switching for the frequency meter

Calibration

Calibration of any frequency meter can be something of a problem. An accurate frequency within the range of the unit is needed, and it should preferably be a frequency that represents about 50 to 100% of the calibration range's full scale value. If your test equipment includes a crystal calibrator, this may well be able to provide a suitable calibration signal. Some calibrators only provide frequencies of about 1MHz and higher, but many provide outputs at 100kHz and lower frequencies. A suitable crystal calibrator design is included in Chapter 2 incidentally.

There are other possible sources for calibration frequencies. Television and radio stations often transmit a test tone (usually at 800Hz or 1kHz) for a while after close down at night. Electronic musical instruments usually have crystal controlled tone generators that have good accuracy. Middle A is at 440Hz, and the A two octaves higher than this is therefore at a frequency of 1.76kHz. This would seem to be ideal for calibrating the unit on the 2kHz range.

Once you have obtained a suitable calibration signal, simply connect it to the input of the frequency meter, switch the unit to the appropriate range, and then adjust VR1 in the d.v.m. module for the correct reading on the display.

Components for Aduio Frequency Meter (Fig.1.11)

Resistors (0.25 watt 5% unless noted)

R1	2M2 1%
R2	220k 1%
R3	22k 1%
R4	2k2 1%
R5	10k
R6	10k
R7	560k
R8	100k
R9	100k
R10	5k6
R11	5k6
R12	100k
R13	5k6

R14	6k8
R15	4k7
R16	1M
R17	1M
Capacitors	
C1	1n polyester
C2	15p ceramic plate
C3	4μ7 63V elect
C4	47n polyester
C5	100μ 16V elect
Semiconductors	
IC1	TLC555CP
IC2	CA3140E
IC3	LF351N
IC4	LF351N
Miscellaneous	
S1	4 way 3 pole rotary (only two poles used)
SK1	3.5mm jack socket
	8 pin d.i.l. i.c. holder (4 off)
	D.V.M. module (12 volt version)

Capacitance Meter

It is possible to produce a good low cost capacitance meter using a circuit similar to the audio frequency meter unit described previously, but with a clock oscillator providing the input signal, and the test capacitor forming the capacitive element in the timing network of the monostable. Rather than having a fixed pulse width and a variable repetition rate, this gives a fixed frequency and a variable pulse width. This gives an average output voltage which varies in sympathy with the test capacitance, and which is linearly proportional to it.

An analogue capacitance meter of this type is described in book number BP248, and it would be quite easy to produce a digital equivalent. The problem with this type of capacitance meter is that stray capacitance in the monostable tends to give poor accuracy when testing very low value capacitors. This stray capacitance is usually about 20p to 80p, and seriously degrades accuracy when measuring capacitances of less than a few hundred picofarads. This is not necessarily too important, since low value capacitors are used much less than medium and high value types. On the other hand, if you are interested in radio you are likely to use large numbers of very low value capacitors, and a unit that can check them properly will then be a decided asset.

It is possible to nullify the built-in capacitance of a capacitance meter that is based on a monostable, but this is more difficult when feeding the output voltage to a digital voltmeter than it is when using a simple moving coil panel meter. Bridge circuits to null offset voltages are easy using a moving coil meter, but not very practical when using a d.v.m. module. After attempts to produce a good monostable based digital capacitance meter design proved fruitless a different approach was tried, and from this the arrangement shown in the block diagram of Figure 1.13 was developed.

This has what could be regarded as a form of clock oscillator, but it is not a clock oscillator of the digital type. Its output is a high quality sinewave signal, and it is important to the accuracy of the unit that this signal is a reasonably pure sinewave type. The next stage is an amplifier having a preset gain control. Perhaps it is not strictly accurate to describe this stage as an amplifier, since its voltage gain is likely to be less than unity. On the other hand, it provides a low output impedance, and could be regarded as an amplifier in this sense. In any event, its primary purpose is to enable the amplitude of the output signal to be varied, and this enables the unit to be calibrated.

The capacitor under test is used to couple the sinewave signal to the next stage, which is an operational amplifier inverting mode circuit. Figure 1.14 shows the standard inverting mode circuit (on the left) and the slightly modified version used in this circuit. The closed loop voltage gain of the standard version is equal to the value of R2 divided by that of R1. In the modified version things are much the same, but the voltage gain is equal to the value of R2 divided by the impedance of C1 (which is the test capacitor). The impedance of a capacitor depends on both signal frequency and the value of the component. In this case the signal frequency is fixed, but

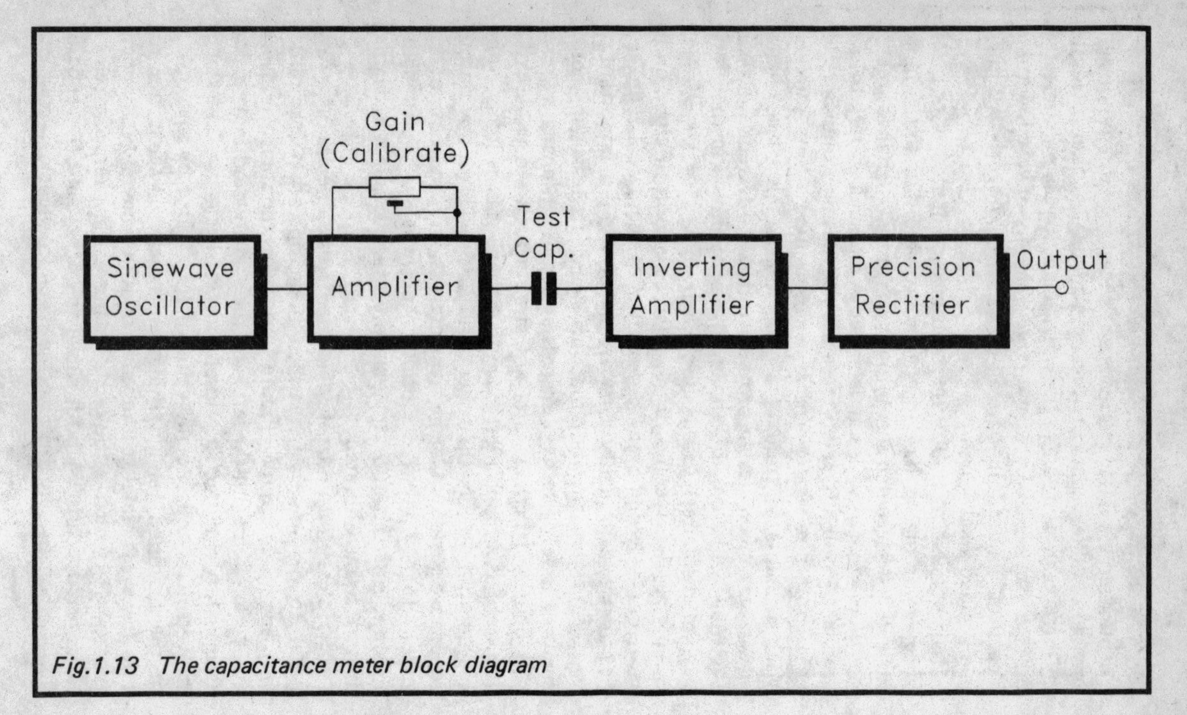

Fig.1.13 The capacitance meter block diagram

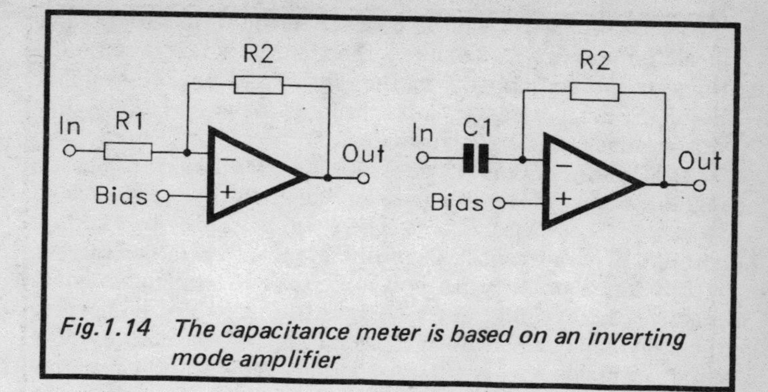

Fig.1.14 The capacitance meter is based on an inverting mode amplifier

the capacitance value is not. The impedance value is inversely proportional to the value of the test component.

As a simple example, assume that R2 has a value of 100k, and that the test capacitance has an impedance of 20k. This means that the amplifier will have a voltage gain of five times (100k/20k = 5). If a capacitor having double the value of the original component is connected to the circuit, its impedance will be only half as high (10k as opposed to 20k). This gives an increase in voltage gain to ten times (100k/10k = 10). If a capacitor having half the value of the original component is connected into the circuit, its impedance will be twice as high, and the voltage gain of the amplifier will be half the original figure (100k/40k = 2.5).

It should be apparent from this that the voltage gain of the amplifier is proportional to the value of the test capacitor. It therefore follows that the output voltage from the amplifier will also be proportional to the value of the test capacitor. In order to obtain a capacitance meter action it is merely necessary to process the output of the amplifier using a precision rectifier, so that the a.c. signal level is converted to a proportional d.c. voltage. This is then fed to the d.v.m. module, and with everything set up correctly the d.v.m. can be made to display the values of test components.

In theory it is not necessary for the signal source to be a highly pure sinewave type. Several component frequencies on the input signal should be satisfactory, since the impedance of the test capacitor will be inversely proportional to its value at all frequencies. In practice the frequency response limitations of the components in the circuit and possible stability problems must be taken into account. A good quality sinewave source ensures good results.

This configuration provides excellent results when measuring low value capacitors as it has no built-in capacitance to swamp the test capacitance. There will inevitably be a certain amount of stray capacitance, but even with a rather careless layout this is unlikely to total more than a few picofarads. In practice there would seem to be little difficulty in keeping the stray capacitance down to under 1p, and this enables quite good accuracy to be obtained even when testing capacitors having values of just a few tens of picofarads. The unit can be used to make reasonably accurate tests on components having values as low as 10p to 20p.

Although it has definite advantages over more simple types of capacitance measuring circuit, this method is not totally free of drawbacks. One is simply that it is a bit more costly and complex than other methods of capacitance measurement. Another is that it is not strictly speaking capacitance that the circuit responds to, but impedance. If you choose a suitable resistance and connect it across the test socket, the unit will respond with a capacitance value for it. In use this could result in misleading results, but it is not likely to do so. A high leakage and non-functioning capacitor could have a resistance that would give a suitable capacitance reading by sheer chance. However, the chances against this happening must be astronomic, and in practice this system gives excellent accuracy, and should give totally reliable results.

The unit described here has five measuring ranges with the following full scale values:–

Range 1 – 1.999n
Range 2 – 19.99n
Range 3 – 199.9n
Range 4 – 1.999µ
Range 5 – 19.99µ

This covers most requirements, including capacitors of just a few picofarads in value. The accuracy when testing these is not particularly good, since they are effectively measured using a single digit display. Results are good enough to sort out the good from the bad though, or to check the value of a component if its markings are not entirely clear. At the high end of the range the unit covers all but high value electrolytic components. However, a simple check with a multimeter will usually reveal whether or not a capacitor of this type is serviceable.

The Circuit

The circuit diagram for the capacitance meter appears in Figure 1.15. The sinewave "clock" signal is generated by IC1, which is used in a Wien oscillator. This is the type of oscillator circuit that is often used in high quality audio signal generators for their sinewave output, and it is very similar to the sinewave generator section of the A.F. signal generator described in book number BP248. In this case though, the Wien network uses only fixed value components since a single output frequency is needed. The output frequency is quite low, and is in fact close to the lower limit of the audio range. Due to the low operating frequency of the circuit, it could be advantageous to make C3 in the d.v.m. module a little higher in value (say 220n).

Th1 provides thermistor stabilisation of the output level. Apart from ensuring that a highly pure output signal is obtained, this also stabilises the output level. Remember that any changes in the output level will be reflected in variations in capacitance readings, and that a constant output level is therefore essential if the unit is to fulfil its potential accuracy. IC2 is the buffer amplifier, and this is a simple inverting mode circuit with VR1 providing both sections of the negative feedback network. VR1 is, of course, the calibration control.

IC3 is the inverting amplifier which has the test capacitance as its input feedback component. S1 is the range switch, and five switched feedback resistors provide the unit with its five measuring ranges. C5 couples the output of IC3 to a conventional precision rectifier circuit. This is a simple half-wave non-inverting type based on IC4. As explained previously, the lowpass filter at the input of the d.v.m. smooths the pulsating

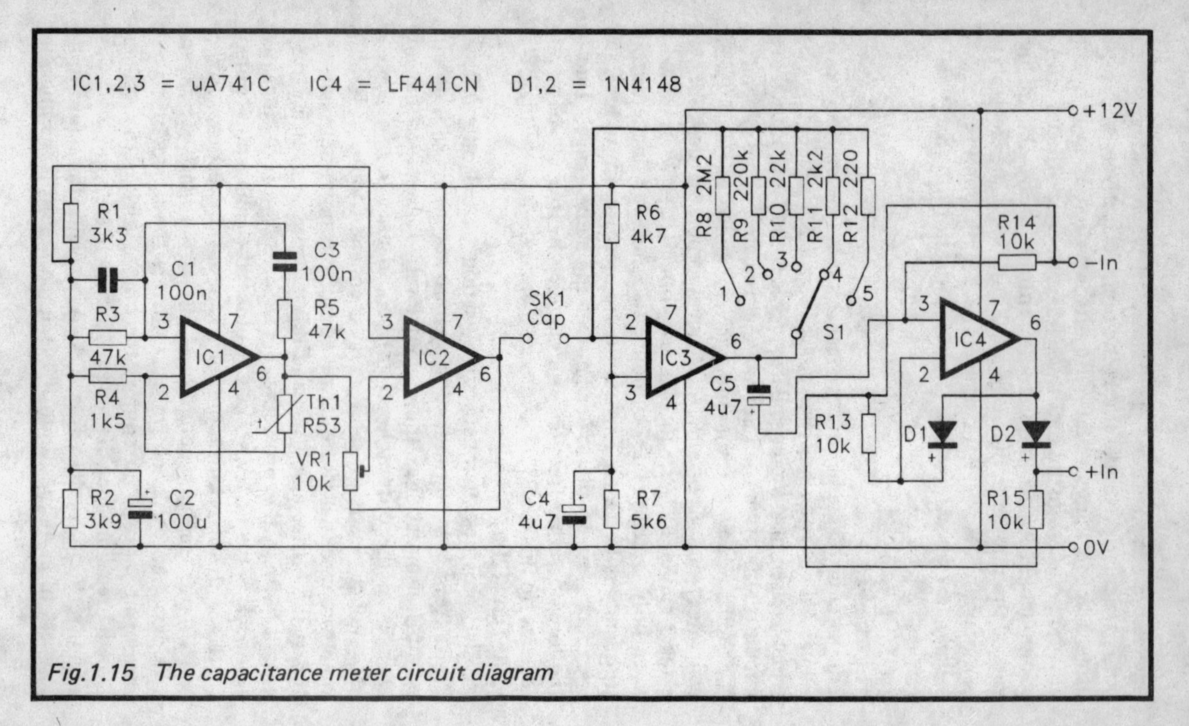

Fig.1.15 The capacitance meter circuit diagram

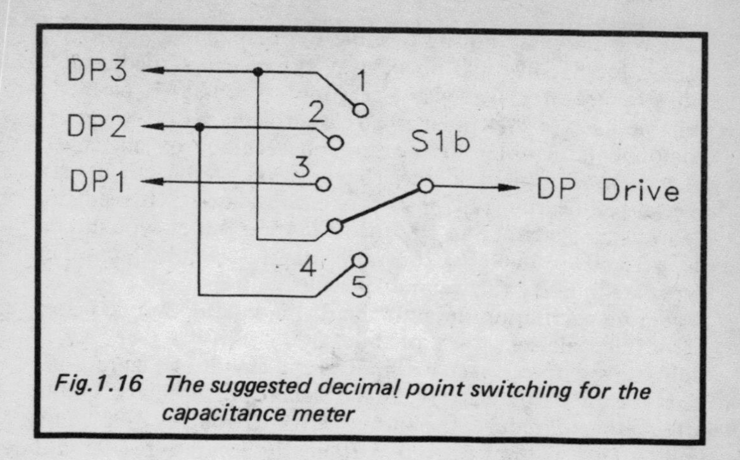

Fig.1.16 The suggested decimal point switching for the capacitance meter

d.c. output of the unit, giving a reasonably ripple-free signal that the voltmeter can measure properly.

Suitable decimal point switching for the capacitance meter unit is shown in Figure 1.16. This provides readings in nanofarads on ranges 1 to 3, and readings in microfarads on ranges 4 and 5.

Calibration

Construction of the unit should not present any major difficulties. The layout is not particularly critical, but you should obviously choose one that puts a minimum of stray capacitance in parallel with SK1. Do not use a screened lead or separate leads twisted together to make the connections between SK1 and the circuit board. Use individual leads kept as far apart as possible. Something like individual 2 millimetre sockets or a twin spring-loaded terminal panel might be a better choice for SK1 than a single two way socket, giving lower stray capacitance. I used a twin spring-loaded terminal unit on the prototype. Many capacitors can be clipped direct to this quite easily. For those that will not connect to it properly a set of test leads must be made up. These should be as short as is practical, so that their self-capacitance is minimised. They are fitted with miniature crocodile clips, which will permit easy connection to any normal type of capacitor.

It will probably not be possible to obtain a five way rotary switch for S1, but a 6 pole 2 way type having an adjustable end-stop (set for five way operation of course) is perfectly suitable. R8 to R12 must all be close tolerance (1% or better) components in order to ensure good accuracy on all ranges. The thermistor used in the Th1 position of the circuit is a special self-heating type in a glass encapsulation. It seems to be available as either an R53 or an RA53. Either type should be perfectly suitable for operation in this circuit, but other types are unlikely to give usable results.

Before calibrating the unit the d.v.m. module should be set for a full scale sensitivity of about 200 millivolts. One way of doing this is to connect its input to the circuit of Figure 1.17. This uses a 9 volt battery as the signal source, plus a simple attenuator to reduce the output level to roughly 200 millivolts. This will not provide particularly accurate results, but it is only necessary for the d.v.m. sensitivity to be roughly 200 millivolts full scale. It is VR1 in the capacitance meter add-on circuit that is used for precise calibration of the unit.

In order to calibrate the unit a close tolerance capacitor is needed, and this should have a value which represents about 50 to 100% of the full scale value of one of the capacitance

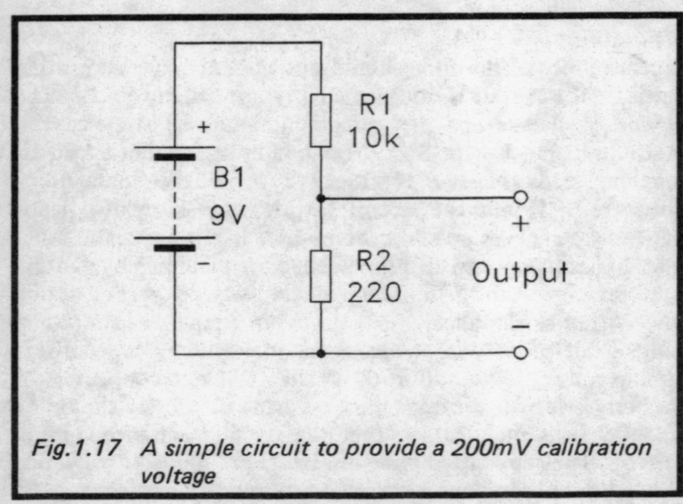

Fig.1.17 A simple circuit to provide a 200mV calibration voltage

meter's ranges. If you do not have a suitable component in the spares box and have to buy one specially, then it is probably best to calibrate the unit on range 1. A 1n or 1.5n 1% capacitor will then suffice, and a close tolerance component at either of these relatively low values should not be very expensive. Simply switch the unit to the appropriate range, connect the test capacitor, and then adjust VR1 in the capacitance meter add-on for the correct reading on the display. The unit should then give accurate results on all five ranges.

When using any capacitance meter there are a few points that should be kept in mind. Firstly, do not connect a charged capacitor to the unit. If the capacitor has only a low voltage charge, then this is unlikely to do any damage. However, charges of more than a few volts, particularly when measuring higher value components, could easily damage the circuit. You may like to add a couple of sockets to the front panel of the unit, with a resistor of about 47 ohms connected between them. Test components can then be connected across these sockets and discharged prior to being measured.

When testing capacitors you should bear in mind that they often have quite wide tolerances. With some of the smaller types the tolerance is around 1% to 5%, but most capacitors have tolerances of 10% or 20%. Certain types commonly have even wider tolerances. In particular, disk ceramic and electrolytic types often have tolerances of about plus 50% and minus 20%, and I have encountered electrolytic capacitors with tolerances of plus 100% and minus 50%. Therefore, if readings sometimes differ quite significantly from the marked values of components, this does not necessarily mean that the components in question are "duds". The discrepancies could be due to the wide tolerances of the test components. When testing close tolerance components, remember that the basic accuracy of the capacitance meter unit is about plus and minus 1% or so, and that an error of about 2% on a 1% component does not necessarily mean that it is outside its specification.

In theory the unit is not suitable for testing electrolytic components since there is no significant d.c. bias across the test socket. In practice there will be a small d.c. bias present here, and connecting electrolytic capacitors one way round or the other should give satisfactory results. A little experimentation should reveal the correct polarity for polarised capacitors (in practice it is quite likely that connecting them with either polarity will give satisfactory results).

Components for Capacitance Meter (Fig.1.15)

Resistors (0.25 watt 5% unless noted)

R1	3k3
R2	3k9
R3	47k
R4	1k5
R5	47k
R6	4k7
R7	5k6
R8	2M2 1%
R9	220k 1%
R10	22k 1%
R11	2k2 1%
R12	220R 1%
R13	10k
R14	10k
R15	10k

Capacitors

C1	100n polyester
C2	100µ 10V elect
C3	100n polyester
C4	4µ7 63V elect
C5	4µ7 63V elect

Semiconductors

IC1	µA741C
IC2	µA741C
IC3	µA741C
IC4	LF441C
D1	1N4148
D2	1N4148

Miscellaneous

S1	5 way 2 pole rotary (6 way with adjustable end-stop)

SK1	See text
Th1	R53 or RA53 thermistor
	8 pin d.i.l. i.c. holder (4 off)
	d.v.m. module (12 volt version)

Resistance Meter

An analogue multimeter on its resistance ranges has a rather inconvenient reverse reading and non-linear scale. This is due to the use of a very simple arrangement which has the meter movement connected in series with the test resistance and a variable resistance which provides electrical zeroing of the meter. This potentiometer is adjusted so that with the test prods short circuited together there is full scale deflection of the meter, and a reading of zero (bearing in mind that the scale is a reverse reading type). Placing a resistance across the test prods gives reduced current flow, and reduced deflection of the meter's pointer. The higher the resistance, the less the deflection of the pointer. Several switched shunt resistors across the meter movement provide several measuring ranges.

This approach has the advantage of extreme simplicity, and although it can be a bit awkward to use at first, most people soon get used to the unusual scaling. This system is totally unusable with a digital readout though. There is no easy digital equivalent to the analogue method of simply fitting a scale to suit the readings obtained. I suppose that it would be possible to use an inverting amplifier and a non-linear type in order to give a linear resistance to voltage conversion, but it would not be worthwhile doing so. There is a much easier way of tackling the problem.

The basic principle on which most digital resistance meters are based is that the voltage developed across a resistor is proportional to its resistance, provided there is a constant current flow. As a few examples, suppose that a constant current of 1 amp is chosen, and that this current is fed through resistors having values of 1, 2, 5, and 10 ohms. Ohm's Law states that voltage is equal to current multiplied by resistance, and it only takes a little mental arithmetic to come up with answers of 1, 2, 5, and 10 volts for our example values. In other words, the required action is obtained with an output voltage that is proportional to the test resistance.

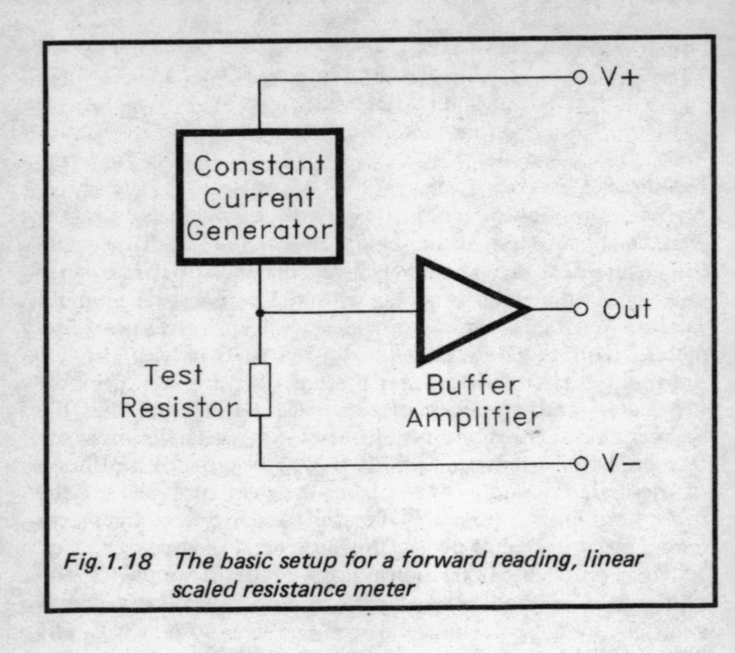

Fig.1.18 The basic setup for a forward reading, linear scaled resistance meter

Therefore, a linear resistance meter interface for a voltmeter only needs to be a simple setup of the type outlined in Figure 1.18. The test resistor is fed from a constant current generator circuit, and if several measuring ranges are needed, this must have several switched output currents. The buffer amplifier ensures that no significant current is tapped off by the voltmeter connected to the output. This prevents the voltmeter circuit from shunting the test resistor, and possibly preventing the system from operating properly. Clearly it would not be possible to measure resistors having values of a few megohms if the voltmeter placed a resistance of a few kilohms across the test points. In this case the buffer amplifier is not required because the d.v.m. module has an exceedingly high input resistance. However, if you should try to use the resistance meter circuit with another voltmeter you might need to add a buffer amplifier in order to obtain satisfactory results.

The Circuit

The circuit diagram for the resistance meter is shown in Figure 1.19. This is basically just a conventional constant current generator circuit of the type which has two diodes to provide a bias of about 1.2 volts to the base of a transistor. Due to the voltage drop of about 0.6 volts from the base to the emitter terminals, this gives about 0.6 volts across the emitter resistor. The collector current of a transistor is virtually identical to the emitter current, and the collector current can therefore be set at the required level by using a suitable emitter resistance. In this case there are five switched emitter resistances, giving the unit five measuring ranges. The full scale values are as follows:–

Range 1	–	199.9 ohms
Range 2	–	1.999k
Range 3	–	19.99k
Range 4	–	199.9k
Range 5	–	1.999M

Very high value resistors are outside the scope of the unit, but these are little used in practice. At the other end of the range the unit is very good, with a resolution of 100 milliohms on the lowest range.

Note that instead of two silicon diodes to give the stabilised 1.2 volt supply at the base of TR1, this circuit actually uses a high quality voltage stabiliser (D1). The problem with a simple diode stabiliser circuit is that it does not give high enough stability. Variations in the battery voltage give significant changes in readings. The 8069 stabiliser is so good that when using this device in the circuit there are no variations in readings even with changes of a few volts in the supply voltage.

A suitable switching arrangement for the decimal point driver is provided in Figure 1.20. This gives readings in ohms on range 1, kilohms on ranges 2, 3, and 4, and megohms on range 5.

Construction of the circuit is very straightforward, and the layout is in no way critical. Adjustment of the preset resistors will be easier if they are multi-turn types, but ordinary

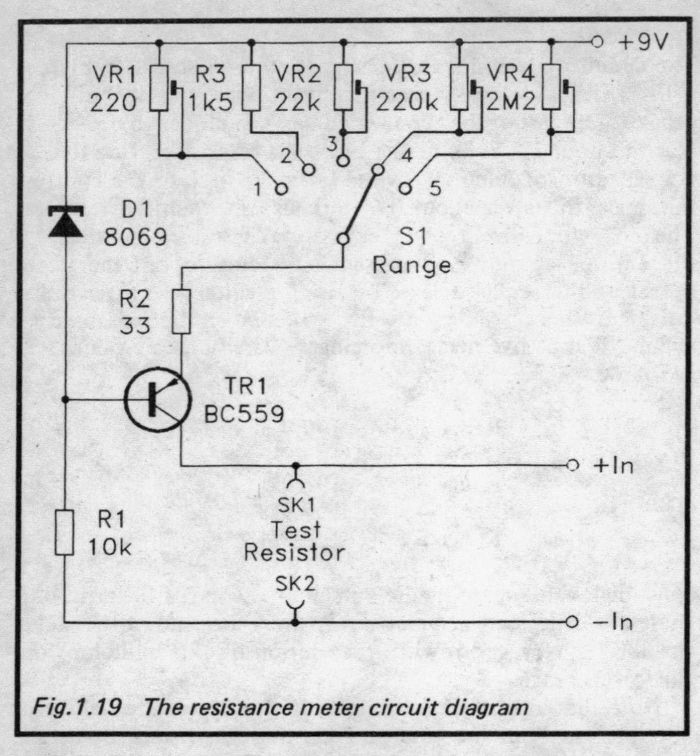

Fig.1.19 The resistance meter circuit diagram

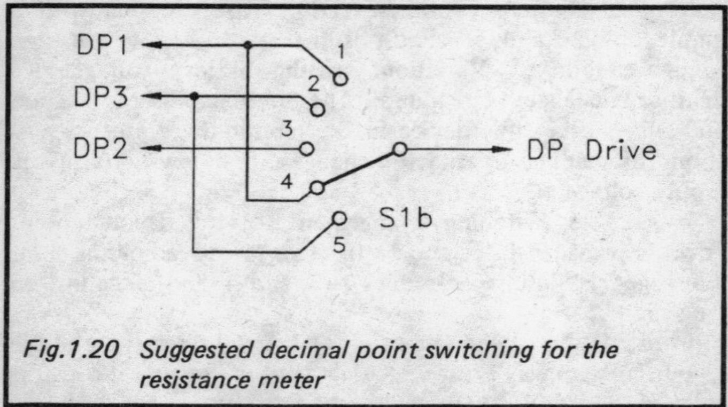

Fig.1.20 Suggested decimal point switching for the resistance meter

miniature presets are just about usable. The unit requires the d.v.m. module to have a somewhat higher full scale input voltage than that needed by the other circuits. It might not be possible to obtain a suitable sensitivity with the original component values, but increasing VR1 from 1k to 10k should ensure that the unit can be calibrated correctly.

Calibration requires five close tolerance resistors having values at something approaching the full scale value of each of the five measuring ranges. The best values to use are therefore 180 ohms, 1k8, 18k, 180k and 1M8 to calibrate ranges 1 to 5 respectively. Close tolerance resistors having values of more than 1 megohm are relatively difficult to obtain and expensive. Accordingly, you might prefer to use a 1M component to calibrate range 3.

Start by switching the unit to range 2 and connecting the 1k8 resistor across SK1 and SK2. Next adjust VR1 in the d.v.m. module for a reading of "1.800". Then connect the 180 ohm resistor across SK1 and SK2, switch the unit to range 1, and adjust VR1 in the resistance meter circuit for a reading of "180.0". Repeat this general procedure using the 18k, 180k and 1M8 resistors to calibrate ranges 3 to 5 respectively, with VR2 to VR4 being used to set the correct readings for their respective ranges. The unit is then ready for use.

Components for Resistance Meter (Fig.1.19)

Resistors (all 0.25 watt 5%)

R1	10k
R2	33R
R3	1k5

Potentiometers

VR1	220R multi-turn preset
VR2	22k multi-turn preset
VR3	220k multi-turn preset
VR4	2M2 multi-turn preset

Semiconductors

D1	8069 1V2 reference voltage generator
TR1	BC559

Miscellaneous

SK1	2mm socket
SK2	2mm socket
S1	5 way 2 pole rotary (6 way 2 pole with adjustable end-stop)
	180R, 1k8, 18k, 180k, and 1M8 1% tolerance resistors for calibration purposes.

Transistor Tester

Comprehensive transistor testers are highly complex pieces of equipment that enable the test parameters (collector voltage, collector current, etc.) to be set at practically any desired levels. For most purposes though, a unit of this type is severely over-specified. Generally what is needed is a device that will give an indication of current gain for both n.p.n. and p.n.p. devices, and which can be used quickly and easily without the need for any complex setting up procedures. The fact that the test conditions are not the same as those specified in the data sheet for the component under test does not matter too much. These differences can be taken into account when assessing results.

Probably the two most important test conditions are the collector current and the collector voltage. In general, the gain of a transistor rises slightly as the collector voltage is increased. The difference is usually quite small even with substantial changes in the collector potential, and this is unlikely to be the cause of significant anomalies provided the collector voltage is not allowed to become very low. The collector current is a different matter though. The gain of a transistor, in the main, rises as its collector current is increased. The difference in gain can be quite dramatic. A transistor that has a current gain of a few hundred times at a collector current of one milliamp might have a gain of less than ten times at a collector current of ten microamps. Differences between the collector current at which a transistor's gain is specified in the data sheet, and the actual current at which it measured, must therefore be taken into account when assessing results.

The basic method of measuring the d.c. current gain of a transistor is shown in Figure 1.21. The arrangement shown in (a) is for n.p.n. transistors, while that of (b) is for p.n.p. types.

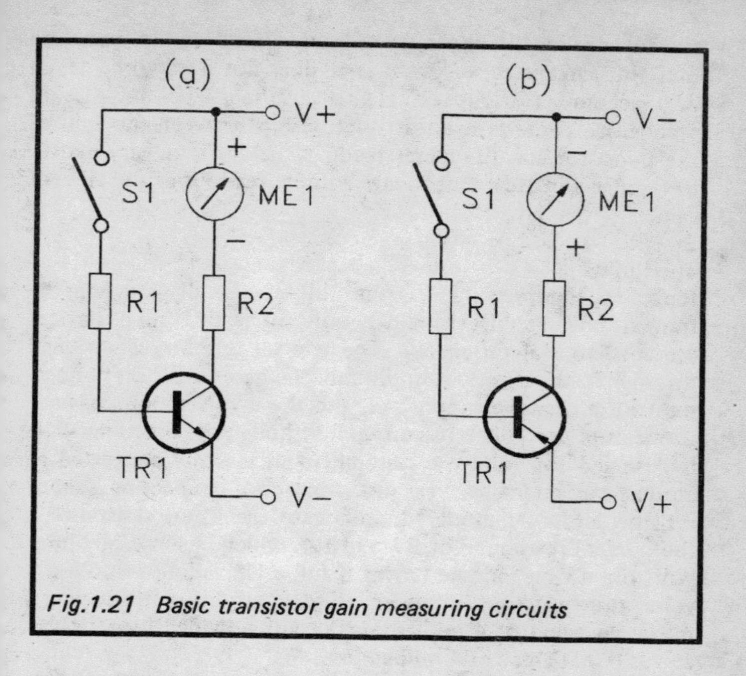

Fig.1.21 Basic transistor gain measuring circuits

They differ only in that the polarity of the supply and the meter. In both cases the transistor is initially tested with S1 open so that no base current is applied to the test device. With an ideal transistor there would be no current flow at all between the collector and emitter terminals, but with practical components at least a small leakage current will flow. For silicon transistors the leakage level will normally be so low to be of no consequence, and in most cases it is so low that it is difficult to measure. The leakage level of germanium transistors is often quite high, making them difficult to test properly. Fortunately, these components are now largely obsolete, and you may never need to test one.

When S1 is closed, a base current is applied to the test component. Resistor R1 limits this base current to the required level. The base current has the effect of causing a larger collector current to flow, and this current is registered on the meter. The current gain of a transistor is equal to the collector current divided by the base current. Here we are assuming that the leakage is so low that it does not need to be taken into account (which will almost invariably be the case in practice). There is a linear relationship between the gain of the transistor and the meter reading, making it an easy matter to arrange for the meter to give a direct readout of the current gain.

The Circuit

Refer to Figure 1.22 for the full circuit diagram of the transistor tester. Devising a circuit for testing p.n.p. devices presents no real problems. The emitter terminal is switched through to the positive supply rail, the base resistor is connected to the negative supply rail, and the d.v.m. module is used to measure the collector current. Although the d.v.m. module is designed for voltage measurement, it is easily converted to current measurement. It is just a matter of connecting a shunt resistor across its input. In this case the shunt resistance is the parallel resistance of R3 and R4, which is some 50 ohms. With the d.v.m. module having a full scale value of 200 millivolts, this effectively converts it to a current meter having a full scale value of 4 milliamps (0.2 volts divided by 50 ohms equals 0.004 amps or 4 milliamps).

Base resistor R5 provides a base current of approximately 2 microamps. Therefore, if the test transistor has a gain of one thousand times it will produce a collector current of 2 milliamps, and will give half full scale value on the d.v.m. module. In other words, a reading of "1000". A current gain of 500 would give a collector current of 1 milliamp, and a 25% of full scale reading from the d.v.m. module (i.e. a reading of "500"). In other words, the circuit provides readings direct in current gain.

Testing p.n.p. transistors poses no problems, since they act as a current source that can directly drive the d.v.m. module. The same is not true of n.p.n. transistors which act as current sinks, and must drive the d.v.m. module via some form of inverter circuit. This inverting action is obtained using a simple current mirror based on TR1 and TR2. The current fed into the base of TR2 produces an almost identical current to be sourced from its collector. For good results the gains of

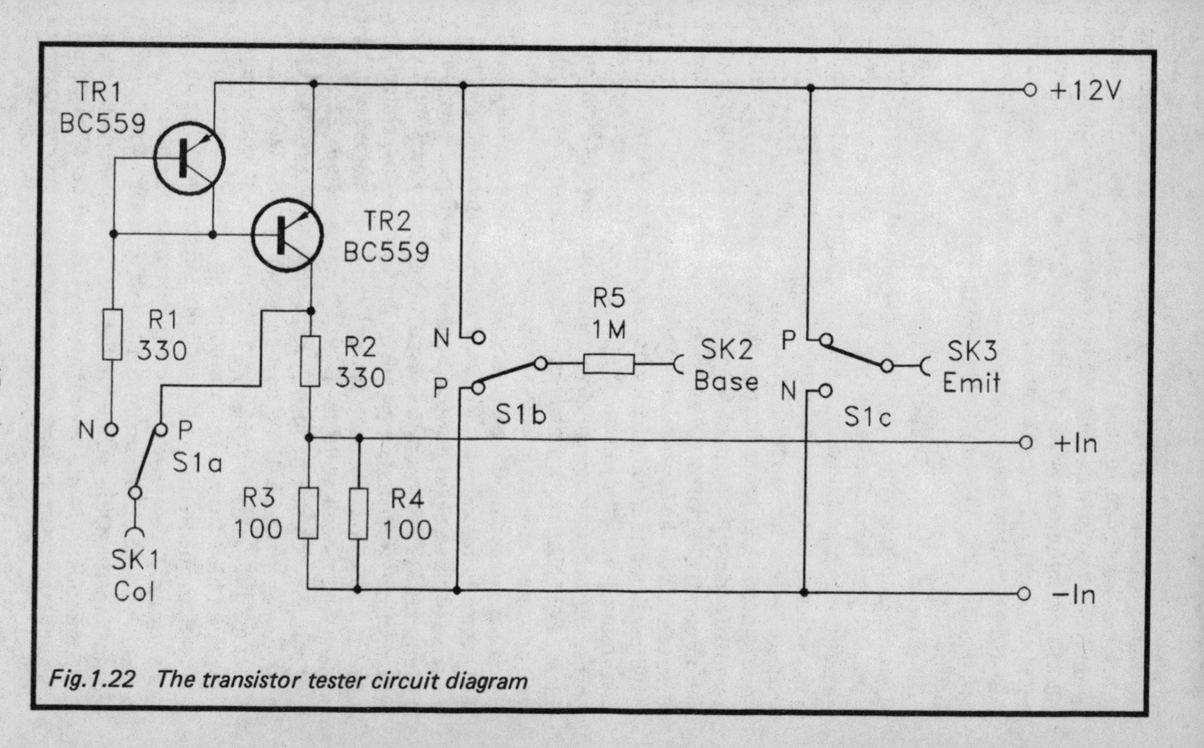

Fig.1.22 The transistor tester circuit diagram

TR1 and TR2 should be reasonably well matched. Obtaining ready-matched transistors could be difficult. One solution to the problem is to buy a few BC559s (which are not expensive), and to initially use any two of them for TR1 and TR2. Then use the tester to check the others so that the two with the most closely matched gains can be sorted out. These are then used as TR1 and TR2, replacing those originally used in the circuit.

The n.p.n./p.n.p. switching is quite straightforward, with S1a switching the collector test socket (SK1) between the input of the d.v.m. module and the input of the current mirror. R1 and R2 are included to protect the unit and devices under test from excessive current flows. There is no need to switch the current mirror out of circuit in the p.n.p. mode, since it will produce only minute leakage currents that are totally inconsequential. S1b switches the base current to the appropriate polarity, while S1c connects the emitter socket (SK3) to the appropriate supply rail.

With a resolution of one the circuit is capable of testing low gain transistors. However, the collector current at which devices are tested becomes very small if they have low current gains. For instance, a transistor with a current gain of 500 will be tested at a collector current of 1 milliamp, but one which has a gain of 50 will be tested at a current of only 100 microamps. This can give what are really artificially low gain readings on lower gain devices, possibly resulting in very misleading results. This is not a major problem if you take into account that low gain devices will be measured at low collector currents. However, you may prefer to add a second measuring range having a full scale value of 199.9 (as opposed to the 1999 of the original circuit). This merely entails adding a d.p.d.t. switch and a 100k resistor, as shown in Figure 1.23. Selecting the 100k resistor instead of the 1M component boosts the base current to approximately 20 microamps. The test devices then only need one-tenth of the current gain in order to produce a given collector current and reading. The second pole of S2 is used to connect the DP drive signal through to the DP1 pin of the liquid crystal display, so that the appropriate decimal point segment is switched on when the additional range is used.

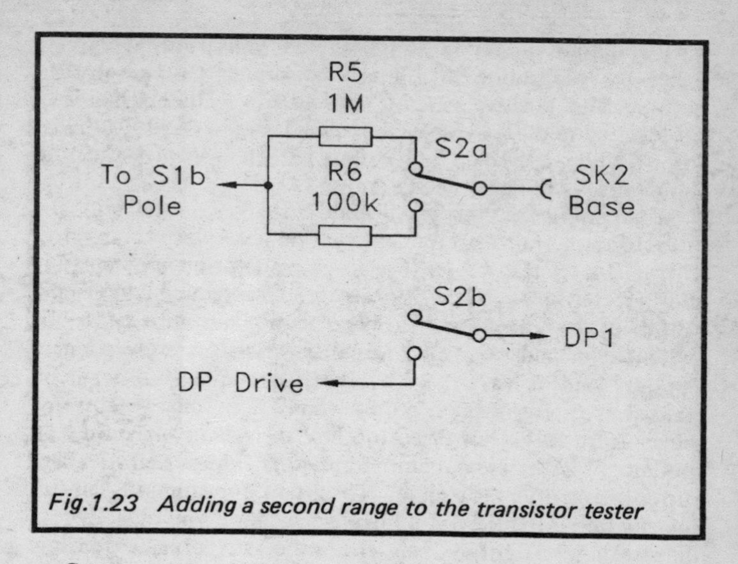

Fig.1.23 Adding a second range to the transistor tester

Construction of the transistor tester should be very simple, and there is nothing critical about the layout. SK1 to SK3 should be miniature (1 or 2 millimetre) sockets mounted close together. Most transistors will then plug directly into them without too much difficulty. Clearly label the sockets and (or) use sockets of different colours to permit easy identification, and reduce the risk of connecting test devices incorrectly (something which is unlikely to damage the transistors or the tester). Some transistors will not easily connect to the sockets, and for these a set of test leads terminated in crocodile clips must be made up. Use leads and (or) clips of different colours so that the leads are easily identified and errors are avoided when connecting test components.

Unlike construction of the unit, calibration does pose something of a problem, but if you have the necessary test equipment it is possible to measure the exact base current and adjust the sensitivity of the d.v.m. module to suit. In practice this would not really seem to be worthwhile. As pointed out previously, results have to be considered subjectively since measurements are not being made at user selected collector currents and voltages. However carefully the unit is calibrated, results will always lack true precision. A simple but adequate means of calibrating the unit is to connect a 1k2 resistor between the positive supply rail and the "+In" terminal of the d.v.m. module. Then adjust VR1 in the d.v.m. module for a reading of "1000". The unit is then ready for use, and should give reasonably accurate results.

When using the unit you should initially connect only the collector and emitter terminals each time you test a transistor. This enables a test for leakage to be made prior to connecting the base terminal and taking a gain measurement. You may prefer to add a switch to permit the base current to be switched on and off, but I found it easier to simply connect the base lead or leave it unconnected, as required. For silicon transistors there should be extremely low leakage currents which in most cases will be too low to register on the d.v.m. module. With germanium transistors quite high leakage currents are often tolerable. To obtain a meaningful current gain figure when there is a high leakage level you must deduct the leakage current reading from the current gain reading.

Components for Transistor Tester (Fig.1.22)

Resistors (all 0.25 watt 5%)

R1	330R
R2	330R
R3	100R
R4	100R
R5	1M

Semiconductors

TR1	BC559
TR2	BC559

Miscellaneous

S1	2 way 3 pole (e.g. 4 way 3 pole rotary switch having an adjustable end-stop set for 2 way operation)
SK1	2mm socket
SK2	2mm socket
SK3	2mm socket
	d.v.m. module (12 volt version)

Current Tracer

A current tracer is a device for testing the current flow through printed circuit tracks and component wires. Normally the measurement of current through either of these requires that the lead or track should be cut so that the current meter can be wired into the signal path. This is obviously time consuming, and rather destructive. Tracks can be repaired, and component leads can be joined together again, but this type of thing can soon reduce a neat circuit board to a piece of junk! Testing of this kind therefore has to be avoided as far as possible, which usually means largely or totally abandoning current checks in favour of voltage tests. This can make life difficult, since detecting an incorrect current flow into a stage of a circuit would often represent the quickest means of detecting the faulty part of a circuit. It can be particularly useful with circuits that contain a number of integrated circuits and few discrete components, as these are often very difficult to test properly using voltage checks.

There are some quite sophisticated pieces of electronics which enable the current flow through a track or wire to be measured quite accurately without actually breaking the circuit and inserting a meter. One system uses a special Hall Effect device to detect the strength of the magnetic field produced by the current flow, and from this the amount of current flow can be determined with reasonable accuracy. As yet, this system is probably not a practical proposition for the home constructor as it is fairly complex, and suitable Hall Effect sensors for this particular application would not seem to be available to the home constructor.

There is a much more simple alternative which is less accurate, but which can provide some useful results. This is really a variation on the standard current measuring technique. Figure 1.24(a) shows the basic setup for current measurement. The current is passed through resistor R1 and meter ME1. The current is split between the two of them, and the lower R1 is made in value, the greater its share of the current flow. R1 therefore enables the full scale current of the circuit to be set at any figure higher than the basic sensitivity of the meter.

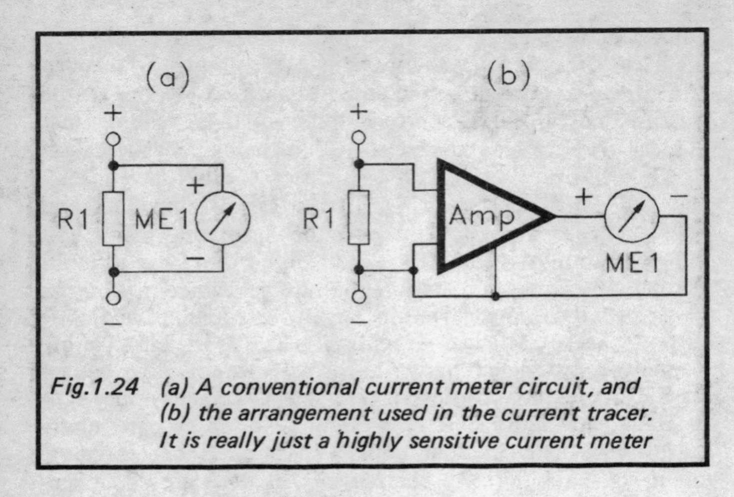

Fig.1.24 (a) A conventional current meter circuit, and (b) the arrangement used in the current tracer. It is really just a highly sensitive current meter

The arrangement shown in Figure 1.24(b) is similar, but a high gain d.c. amplifier has been added between the shunt resistor and the voltmeter circuit. This has the effect of making the circuit extremely sensitive, being able to readily detect currents into the nanoamp region. If the input of the circuit is connected across a short piece of track or a length of leadout wire, some of the current flowing in that wire or track will flow through R1 and will be registered on the meter. With the leadout wire or track having a very low resistance in comparison to R1 the proportion of the total current flow that will be redirected through R1 is likely to be extremely small. However, provided the amplifier is sufficiently sensitive, this current will be detected and indicated by the meter. The percentage of current that the unit redirects will remain the same regardless of the actual current flow through the wire or track. Therefore, the greater the current flow, the greater the reading on the meter.

This method lacks precision in that the proportion of current that the tracer unit will redirect is dependent on several factors. One is simply the distance between the test prods. The greater this distance, the higher the track or lead resistance between them, and the greater the proportion of current that the unit will tap off. Using a fixed amount of separation can help to give more meaningful results, but does not totally solve the problem. The resistance per millimetre of track or lead will vary, and will be more for thin leads or tracks. The quality of the connection between the test prods and the test points can also have a significant effect on readings.

Consequently, this method of current tracing does not give highly accurate results. It does give a good guide to the amount of current going through a track though. Often a fault will result in a stage consuming a current several times higher than its normal level, or only a fraction of its normal level. These large discrepancies are readily detected using even a simple current tracer of this type. It is the type of thing where experience counts for a great deal, and it is a good idea to use the unit to prod around on some functioning circuits, getting used to the sorts of readings that various levels of current flow produce. You will then quickly spot any readings that are inappropriate for the correct current flow.

The Circuit

Figure 1.25 shows the circuit diagram for the current tracer. This is just a standard operational amplifier inverting mode circuit. The closed loop voltage gain is set by R1 and R2 at a little over 46dB (200 times), and together with the low full scale value of the d.v.m. circuit, this gives excellent sensitivity.

VR1 is an offset null control. With a theoretically perfect operational amplifier this is unnecessary, but with practical devices the input and output voltages can drift away from the correct levels. This problem can be quite acute in a circuit of this type which has quite a high d.c. voltage gain. In fact the input offsets when multiplied by the closed loop voltage gain of the amplifier can be so high that the output of the amplifier simply goes fully negative or fully positive under quiescent conditions! VR1 enables any offset voltages to be trimmed out so that zero output voltage is produced under standby conditions.

The unit should not prove to be difficult to construct. IC1 has quite a high voltage gain, but due to its low input impedance and the fact that its input and output are out-of-phase, instability is unlikely to occur. It is advisable to use a multi-

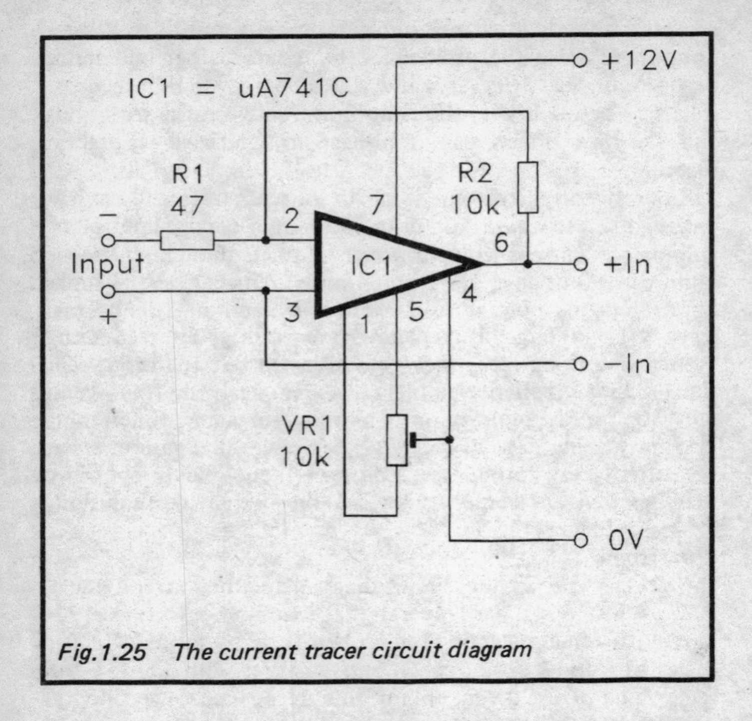

Fig.1.25 The current tracer circuit diagram

turn preset for VR1. Adjustment of this control is very critical, and it might not be possible to accurately trim out the offset voltages if an ordinary preset resistor is used in the VR1 position. When the unit is first switched on it is unlikely that any offsets will be apparent. The reason for this is that IC1 is used in the inverting mode, and there is effectively 100% negative feedback through R2 (and only unity voltage gain) when the input is open circuit. Short circuiting the test prods together will almost certainly produce a large change in the output voltage, probably resulting in an overload indication from the d.v.m. module. By carefully adjusting VR1 it should be possible to trim the output voltage down to zero, and it should remain at zero when the short circuit across the test prods is removed. Note that with this particular project the setting of VR1 in the d.v.m. module is not particularly important. Any setting that gives good results can be used, and virtually any setting of VR1 will probably fall in this category.

As pointed out previously, interpreting results with a unit of this type is something that can only be based on practical experience. You therefore need to experiment a little with the unit before trying to use it in earnest. It will detect quite small current flows, but if there is only a microamp or two flowing through a printed circuit track, it would be unrealistic to expect a unit of this type to detect it. The polarity indicator on the display will show the polarity of the input signal, but bear in mind that the unit can not detect a.c. current flows. It detects the average current flow if the input signal is not a constant d.c. type. The average current in an a.c. signal is zero, with the positive half cycles cancelling out the negative ones. This is not a major drawback in practice since few circuits contain true a.c. current flows. They are mostly constant or pulsing d.c. types, both of which the current tracer unit can detect properly.

Components for Current Tracer (Fig.1.25)

Resistors (all 0.25 watt 5%)

R1	47R
R2	10k

Potentiometer

VR1	10k multi-turn preset

Semiconductor

IC1	μA741C

Miscellaneous

8 pin d.i.l. i.c. holder
d.v.m. module (12 volt version)

Heatsink Thermometer

When dealing with semiconductors operating at high power levels there is the ever present danger of overheating occurring. If the thermal conduction between a semiconductor and its

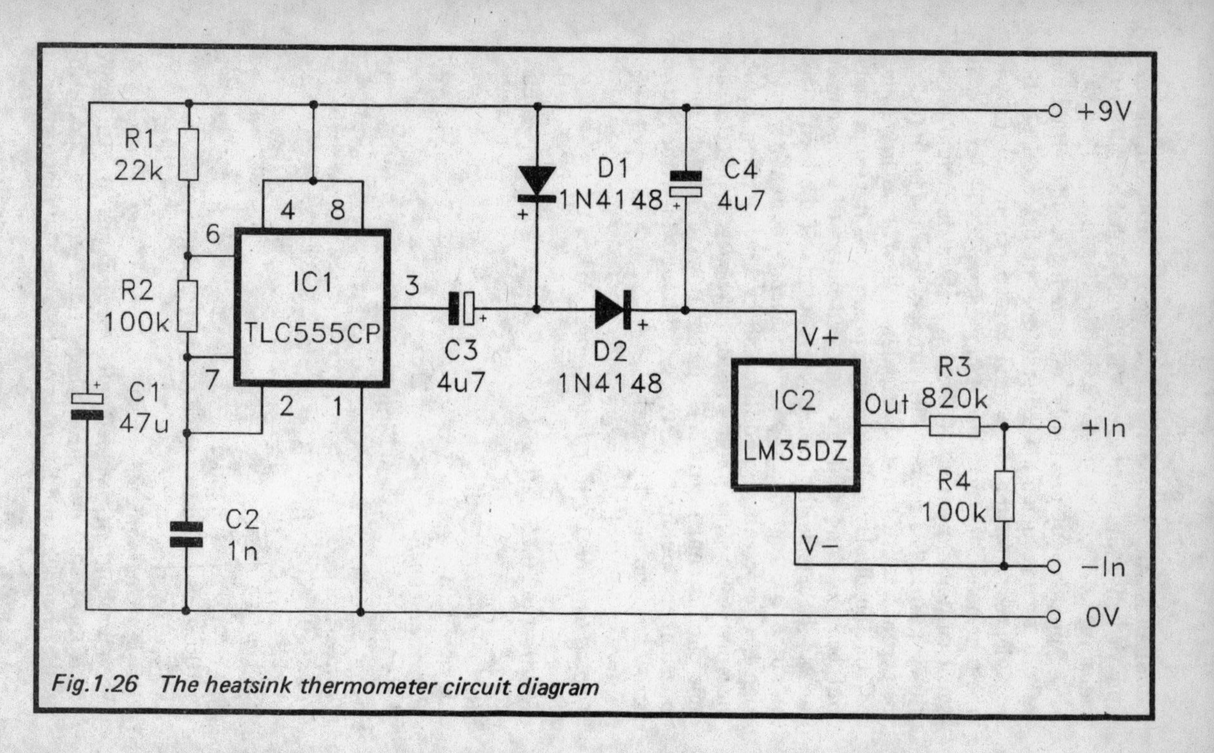

Fig.1.26 The heatsink thermometer circuit diagram

heatsink is not all it could be, you are not likely to become aware of the fact until the device becomes excessively hot and breaks down. This can be spectacular, and more than slightly dangerous. Overheating semiconductors have a nasty habit of exploding with a loud "crack", with bits of plastic, etc. flying in all directions!

A thermometer can be more than a little helpful when checking that power circuits are operating within their designed temperature limits. You can check that devices are operating at suitable temperatures, and do something about it before it is too late if an excessive temperature is detected. By measuring the difference between the case and heatsink temperatures you can get a good idea of how efficiently (or otherwise) the power device is transferring its heat to the heatsink.

Obtaining a thermometer action from the d.v.m. module is quite simple, either using a discrete temperature sensor based on a diode sensor, or using a special integrated circuit which also utilizes what is essentially a diode sensor. The basic method of using a diode as a temperature sensor is to forward bias it from a constant current source. The voltage developed across a forward biased silicon junction is about 0.65 volts, but this figure varies with temperature. The change is a reduction of only about 2 or 3 millivolts per degree Celsius increase, but this is sufficient for many purposes and can easily be amplified if necessary. The advantage of semiconductor sensors over other simple forms of temperature sensor is that they provide good linearity over a wide temperature range.

The circuit (Figure 1.26) of this thermometer is based on a special temperature sensor integrated circuit, the LM35DZ (IC2). This provides an output voltage equal to 10 millivolts per degree Celsius over the range 0 to 100 degrees Celsius. This makes it well suited to the current application where the temperature range of main interest is around 40 to 100 degrees Celsius. Accuracy at low temperatures is not very good, but this is clearly of no consequence in the current application. The output of IC2 is about ten times higher than is needed to drive the d.v.m. module, and so R3 and R4 are used to attenuate its output to a suitable level. The DP drive signal should be connected through to the "DP1" segment of the liquid crystal display. The d.v.m. module can then be set-up to give readings from 0 to 100.0 degrees Celsius. Although the display can go as high as 199.9 degrees Celsius, the accuracy of readings over 100 degrees can not be guaranteed, and there is a risk of the sensor being damaged by overheating.

One slight problem is that the LM35DZ requires a minimum supply voltage of 4 volts, but the potential from the "–In" terminal to the positive supply rail is a volt or so less than this. IC1 is therefore used as an oscillator driving a smoothing and rectifier circuit which provides a potential of about +8 volts referenced to the positive supply rail. This is used as the positive supply rail for IC2, and gives it a total supply voltage of about 10 volts. Changes in battery voltage result in significant variations in the supply voltage to the LM35DZ. This does not matter though, as this device has built-in stabiliser circuits which render changes in the supply voltage innocuous.

Construction presents nothing out of the ordinary apart from the fact that the sensor is not mounted on the circuit board. Instead it is connected to the rest of the unit via a three way lead about half a metre or so in length. Ideally the cable should be a twin screened type with the outer braiding carrying the connection to the "V–" terminal of IC2. With a little initiative it should be possible to make the sensor into a neat probe assembly. This is not essential though, and the main point to watch is that IC2's leads are properly insulated from one another so that short circuits are avoided.

Calibration requires the sensor to be subjected to a known temperature. VR1 in the d.v.m. module is then adjusted to produce the appropriate reading on the display. In order to calibrate the unit you will need a thermometer against which it can be set up. There is no need for any elaborate procedures. Measure the room temperature using the calibration thermometer, and then simply adjust VR1 in the d.v.m. module to give the same reading on the liquid crystal display. It obviously helps if the calibration thermometer is a good quality type which offers good accuracy. It is also preferable if the unit is calibrated at a reasonably high temperature. One at about 20 to 25 degrees is better than about 10 to 15 degrees, where the unit is likely to offer slightly lower accuracy.

When using the unit bear in mind that it takes at least a few seconds for the sensor to adjust to new temperatures. In fact it could take as much as a minute or two for it to fully adjust to very large changes in temperature.

Components for Heatsink Thermometer (Fig. 1.26)

Resistors (all 0.25 watt 5%)

R1	22k
R2	100k
R3	820k
R4	100k

Capacitors

C1	47µ 16V elect
C2	1n polyester
C3	4µ7 63V elect
C4	4µ7 63V elect

Semiconductors

IC1	TLC555CP
IC2	LM35DZ
D1	1N4148
D2	1N4148

Miscellaneous

8 pin d.i.l. i.c. holder
Twin screened lead

Mains PSU

The projects described in this chapter are all suitable for battery operation, but they can obviously be built as mains powered units if preferred. The mains power supply circuit of Figure 1.27 is suitable as the power source for all the projects featured in this chapter. It has full wave bridge rectification, smoothing provided by C1, and stabilisation and further smoothing provided by IC1. It provides a well stabilised 12 volt output. Note that where a 9 volt supply has been

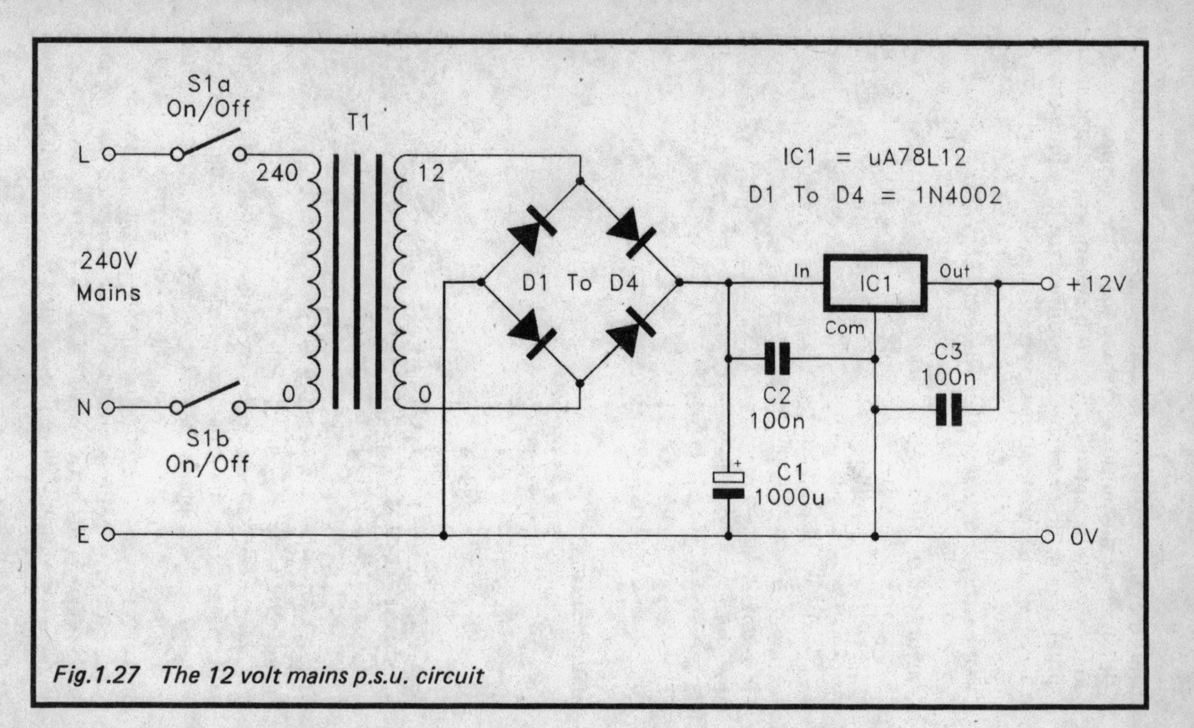

Fig.1.27 The 12 volt mains p.s.u. circuit

specified for projects it is quite in order to use a 12 volt supply.

Construction of the power supply is in most respects quite simple, but as the mains supply is involved it is important to take due care. Mistakes could prove costly and are potentially very dangerous. The unit must be housed in a case which has a screw fitting lid rather than a lid or cover which simply unclips. This prevents easy access to the dangerous mains wiring. Any exposed metalwork must be earthed to the mains earth lead, and this includes such things as fixing screws and sockets. In practice the best solution to this problem is usually to use a case of all metal construction. If the case is earthed to the mains earth lead, then any screws, etc. fitted on it will be earthed as well. C2 and C3 are needed to aid the stability of IC1, and they should be mounted as close to IC1 as reasonably possible.

Components for 12V Mains PSU (Fig. 1.27)

Capacitors

C1	1000μ 25V elect
C2	100n ceramic
C3	100n ceramic

Semiconductors

IC1	μA78L12 (12 volt 100mA positive regulator)
D1	1N4002
D2	1N4002
D3	1N4002
D4	1N4002

Miscellaneous

T1	Mains primary, 12 volt 100mA secondary
S1	Rotary mains switch
	Mains lead and plug

Chapter 2

MISCELLANEOUS TEST GEAR PROJECTS

About the only feature that the projects described in this chapter have in common is that they all make very useful additions to the electronics workshop. Whereas the projects in the previous chapter were concerned with measurements, the projects in this chapter are mainly concerned with the generation of signals of one kind or another. The exception is the final project, which is an unusual but useful form of transistor tester.

Crystal Calibrator

For an electronics hobbyist who is interested in radio, particularly short wave radio, an r.f. signal generator is a more than slightly useful piece of test equipment. A unit of this type does not need to be particularly complex, and is basically just a multi-range r.f. oscillator plus a buffer amplifier. There is a severe difficulty with a home constructed r.f. signal generator, which is the problem of getting the finished unit accurately calibrated. Even if you have access to suitable test equipment, such as a d.f.m. or an accurately calibrated radio that covers the appropriate frequency range, marking large numbers of calibration points on the dial with adequate accuracy is a difficult and extremely time consuming business.

A more simple alternative is to opt for a crystal calibrator. This can not replace a signal generator in every situation, but it will often do what is required. In some cases it will provide better results than a simple r.f. signal generator, since its frequency accuracy and stability are likely to be vastly superior to a simple r.f. generator design. In its most simple form a crystal calibrator merely consists of a crystal oscillator operating at a useful calibration frequency. This would typically be 1MHz. Although a unit which provides an output at just one frequency, albeit with great accuracy, might not seem to be particularly useful, the salient point here is that the unit is designed to generate strong harmonics at frequencies up to and beyond the 30MHz upper limit of the short wave range. Although the basic output frequency may be 1MHz, there are also signals at 2MHz, 3MHz, 4MHz, 5MHz, etc.

Modern crystal calibrators almost invariably offer a number of fundamental output frequencies. The unit featured here offers outputs at 10MHz, 5MHz, 1MHz, 100kHz, and 10kHz. On the face of it, the higher frequencies are unnecessary as they only provide signals at frequencies that are also present on the lower frequency outputs. For instance, the second harmonic of 10MHz is 20MHz, but a 20MHz signal is also provided by the fourth harmonic of the 5MHz output, the twentieth harmonic of the 1MHz output, the two hundredth harmonic of the 100kHz output, and the two thousandth harmonic of the 10kHz output.

In practice matters are not as simple and straightforward as this. You may well be able to obtain a 20MHz signal from the 10kHz output, but with signals at just 10kHz intervals it would probably be impossible to tell which harmonic is which. This problem can be solved by using the 10MHz signal to give the required 20MHz signal. With the harmonics spaced some 10MHz apart there should be no difficulty in sorting out one harmonic from another, even when testing an uncalibrated receiver.

Suppose that you do not require a frequency as convenient as 10MHz. A frequency such as 11.21MHz would seem to be difficult to locate accurately. In practice a frequency of this type is quite easy to set on the receiver. First use the 10MHz signal and locate it on the receiver. Then switch to the 1MHz signal and tune the receiver higher in frequency until a harmonic is located. This signal will be the first 1MHz harmonic at a frequency above 10MHz, and will obviously be the eleventh harmonic at 11MHz. Next switch to the 100kHz signal, and tune the receiver higher in frequency, going through the first signal (11.1MHz) and on to the next one, which will be at 11.2MHz. Finally, switch to the 10kHz signal, and tune the receiver higher in frequency until the next harmonic is located. This will be at the desired frequency of 11.21MHz.

Using this system it is possible to accurately locate any required frequency that is a multiple of 10kHz. It may take a little while to get there, but provided the procedure is carried out carefully, excellent accuracy can be obtained. It is only fair to point out that some receivers have quite wide bandwidths, and it may be difficult to use the unit on its 10kHz output with a set of this type. However, most receivers can use the 10kHz output without any difficulty, and even with a wide bandwidth set the 10kHz output should just about be usable.

The Circuit

In days gone by, the standard method of obtaining several output frequencies from a crystal calibrator was to have several crystal oscillators, with switching to select the output from the desired oscillator. This was an extremely expensive way of tackling the problem, bearing in mind the extremely high cost of crystals in those days. Crystals cost very much less now, but it is still cheaper to use just a single oscillator operating at a high frequency, and feeding a series of frequency divider stages. The design featured here uses the arrangement shown in the block diagram of Figure 2.1.

This has a 10MHz oscillator to generate the basic signal, followed by a buffer amplifier which enables it to drive the next stage properly. This is a divide by two stage which gives a 5MHz output, and this is followed by a divide by 5 type which gives a 1MHz output. The 1MHz signal is then passed to a divide by 10 circuit, followed by another divide by 10 stage, giving outputs at 100kHz and 10kHz. A switch selects the required output and passes it through to a monostable circuit. The output signals from the oscillator and divider stages are reasonably accurate squarewaves having a 1 to 1 mark-space ratio. The problem with a signal of this type is that the even order harmonics are relatively weak. In fact, in theory at any rate, they are totally absent. In practice the accuracy of the waveform is unlikely to be good enough to leave many harmonics undetectable, but the extreme inequality in the strengths of the odd and even order harmonics could make the unit difficult to use, and could easily lead to errors.

This problem is overcome by using the monostable to produce a very brief output pulse on each input cycle. This gives what are relatively low level harmonics at low frequencies,

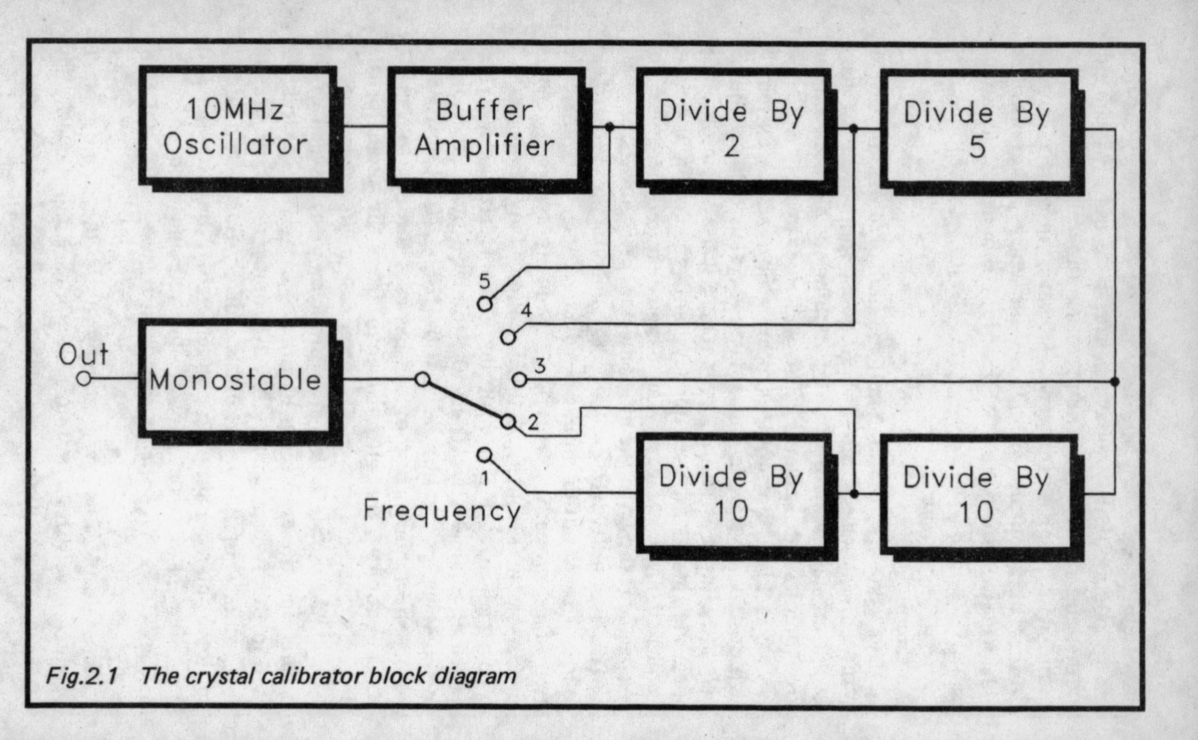

Fig.2.1 The crystal calibrator block diagram

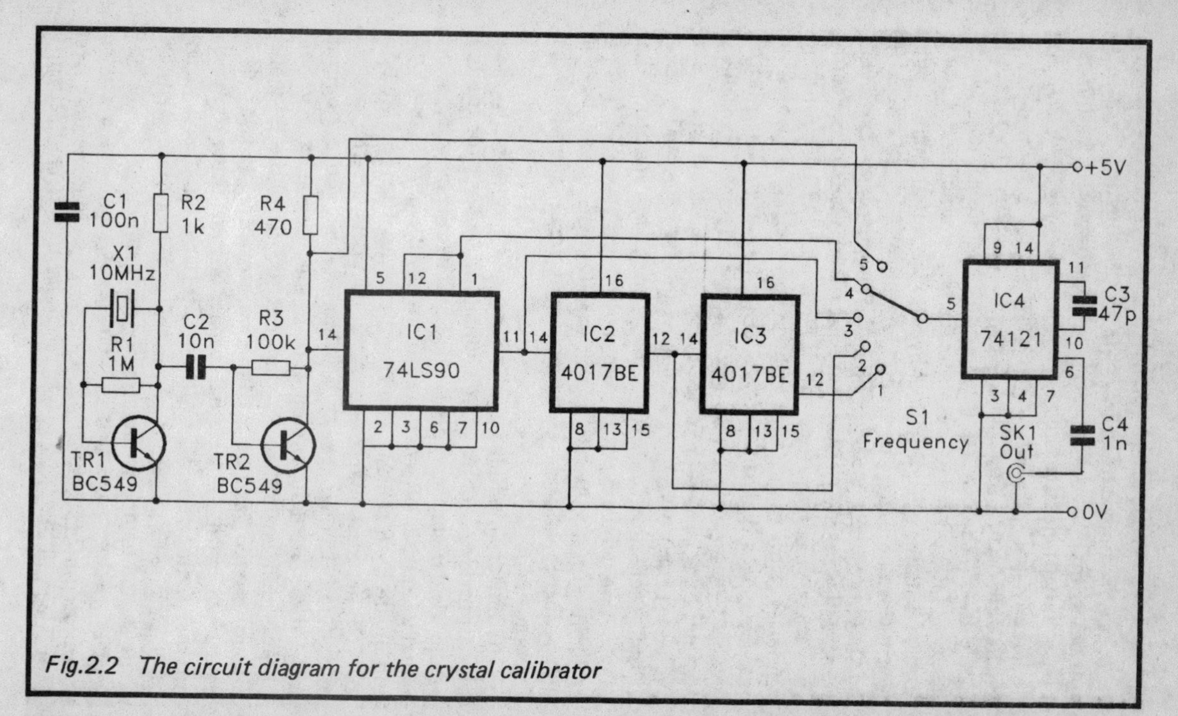

Fig.2.2 The circuit diagram for the crystal calibrator

but it also has the effect of minimising problems with very weak or missing harmonics. It is the higher frequency harmonics that are of primary interest anyway, and these should be sufficiently strong to be readily detectable by any moderately sensitive short wave receiver.

The full circuit diagram of the crystal calibrator appears in Figure 2.2. The crystal oscillator is a conventional design based on TR1, with TR2 acting as the buffer amplifier. Actually, the oscillator is slightly unconventional in that there should be a low value capacitor from the base of TR1 to the earth rail, and another one from TR1's collector to the earth rail. These will almost certainly be unnecessary due to the fairly high operating frequency of the circuit. This results in extremely low capacitance values being needed, which in practice means that stray circuit capacitances are likely to be adequate. However, if TR1 seems to be reluctant to oscillate it would be a good idea to try adding these capacitors. Ceramic plate capacitors having a value of about 6p8 should suffice. Crystal oscillators sometimes have trimmer capacitors that enable fine adjustment of the output frequency. This is probably not worthwhile in this case though, as any error in the frequency is likely to be extremely small, and of no real consequence.

The buffer amplifier is just a simple common emitter stage which provides a more than adequate drive level for the first of the divider stages. This is a 74LS90 decade counter, which actually consists of what are effectively separate divide by 2 and divide by 5 stages. The two divide by 10 stages are both 4017BE decade counter/one of ten decoders. In this case the one of ten decoder facility is of no interest, and the outputs from these are left unused. It is the normal "carry out" outputs that are utilized in this circuit. S1 is the frequency selector switch, and IC4 is the monostable. It is a 74121 connected with an external capacitor but using only the internal timing resistor. Although this circuit uses a variety of logic circuit types, it nevertheless works quite well, and there should be no incompatibility problems.

The circuit requires a 5 volt supply and has a current consumption of about 30 milliamps. Powering the unit from four 1.5 volt batteries wired in series would give a slightly excessive supply voltage and is not recommended. Four NiCad rechargeable batteries are a better choice as they have a potential of about 1.25 volts each, giving the required 5 volt supply voltage. An alternative is to use a fairly high capacity 9 volt battery, such as a PP9 type or six HP7 size cells in a plastic holder, plus a 5 volt regulator to give the required supply voltage. A suitable circuit is provided in Figure 2.3.

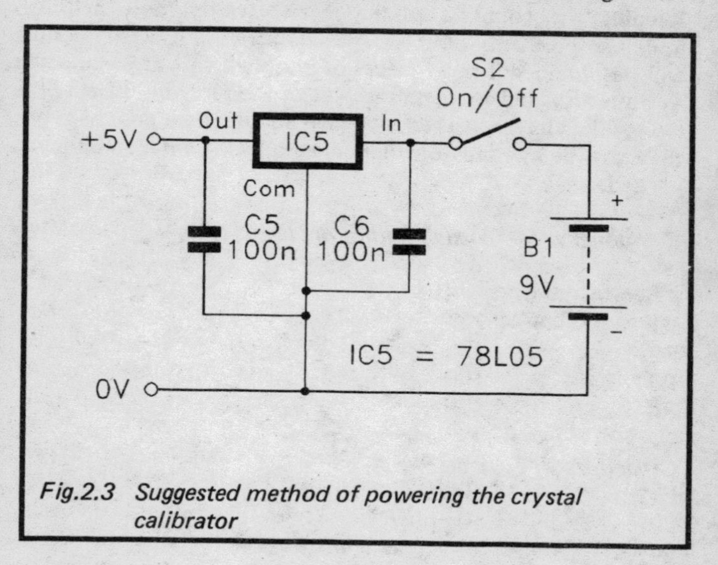

Fig.2.3 Suggested method of powering the crystal calibrator

Construction of the unit is perfectly straightforward, but remember that IC2 and IC3 are CMOS components, and that they consequently require the usual anti-static handling precautions to be observed. S1 can be one pole of a six way two pole rotary switch having an adjustable end-stop set for five way operation. This should be a break before make switch and not a make before break type. The latter would momentarily short circuit two outputs together each time it was set to a new position. This might not damage any of the integrated circuits, but it is obviously better to avoid the risk.

In use it should not be necessary to connect SK1 direct to the aerial socket of the receiver in order to get an adequate signal coupling. In fact it is inadvisable to do this as it would almost certainly provide an excessive input signal. Most receivers have spurious responses, and the nature of a crystal calibrator is such that it provides a large number of output frequencies. This can result in misleading results with what appears to be the desired harmonic actually being a different harmonic picked up on a spurious response of the receiver. Keeping the coupling to the receiver fairly loose helps to minimise problems of this type. Usually satisfactory results will be obtained if short pieces of insulated wire are connected to both SK1 and the aerial socket of the receiver. Placing the two wires close together may give an adequate coupling, but they can be twisted together if a slightly tighter coupling is needed.

Components for Crystal Calibrator (Fig.2.2)

Resistors (all 0.25 watt 5%)

R1	1M
R2	1k
R3	100k
R4	470R

Capacitors

C1	100n ceramic
C2	10n polyester
C3	47p ceramic plate
C4	1n polyester

Semiconductors

IC1	74LS90
IC2	4017BE
IC3	4017BE
IC4	74121
TR1	BC549
TR2	BC549

Miscellaneous

SK1	Coax socket
S1	5 way 1 pole rotary (see text)

X1	10MHz miniature wire-ended crystal
	14 pin d.i.l. i.c. holder (2 off)
	16 pin d.i.l. i.c. holder (2 off)
	Power source (see text)

Additional Components for Power Supply (Fig.2.3)

Capacitors

C5	100n ceramic
C6	100n ceramic

Semiconductor

IC5	78L05

Miscellaneous

S2	s.p.s.t. switch
B1	9 volt (PP9 size)
	Battery clips

Bench Power Supply

A bench power supply must rate as one of the most useful pieces of equipment for the electronics workshop. The unit featured here provides output voltages from a little under 3 volts to 20 volts at currents of up to 1.5 amps. In fact output voltages of up to about 25 volts are available if desired, but above about 20 volts the full rated output current of 1.5 amps can not be achieved. Output overload protection is included in the form of output current limiting. There are three switched limit levels of 15mA, 150mA, and 1.5A. The unit is adequate for most requirements, and it can be used when testing and developing 5 volt logic circuits, 9 volt battery powered equipment, small audio power amplifiers, etc.

The unit offers quite a high level of performance, but the use of a high quality voltage regulator integrated circuit enables quite a simple circuit to be used. The voltage regulator is based on the standard configuration of Figure 2.4. An operational amplifier is at the heart of the unit, and this operates in what is really just a variation on the standard non-inverting amplifier mode. It differs from the standard mode in that the operational amplifier has its output stage augmented by a high power emitter follower output stage (which is

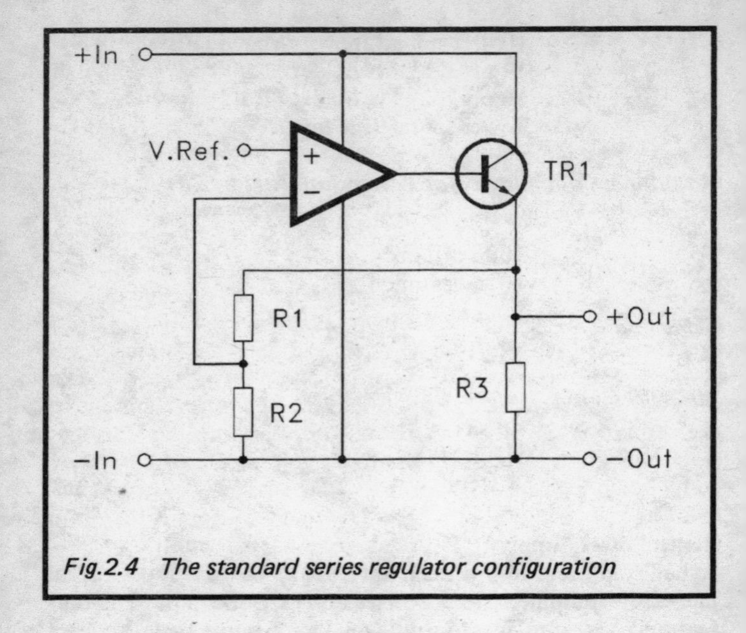

Fig.2.4 The standard series regulator configuration

usually a Darlington Pair in practical circuits). R1 and R2 are the usual negative feedback network which control the closed loop gain of the amplifier. R3 is a load resistor for the output stage, and this merely ensures that there is sufficient load across the output to enable the output stage to function properly. In most cases the current flow through the feedback network is sufficient and no additional loading is required. The non-inverting input of the amplifier is fed with a highly stable reference voltage.

The action of the circuit is to amplify the reference voltage by an amount that is equal to the closed loop gain of the amplifier. The minimum closed loop gain for a non-inverting mode amplifier is unity, and this occurs when the output and inverting input are connected together so that there is 100% negative feedback. This is an important point, since it means that the minimum output voltage of the circuit is equal to the reference voltage. The reference voltage must therefore be made quite low if a reasonably low minimum output potential is to be achieved. In most cases it is at about 1 to 3 volts.

The closed loop voltage gain of the amplifier is equal to (R1 + R2) divided by R2, and quite high levels of voltage gain are easily achieved. In theory this enables quite high output voltages to be achieved, but in practice matters are not quite as simple as this. The voltage ratings of the semiconductors in the circuit will place a strict limit on the maximum permissible output voltage. Also, as the output voltage is raised, regulation efficiency reduces. Note that the feedback is taken via the output stage and not from the output of the operational amplifier. The feedback therefore compensates for any voltage drops through the output stage. In fact the feedback will attempt to compensate for any changes in output voltage, such as those caused by variations in loading, fluctuations in the mains voltage, etc. It will not compensate for any changes caused by variations in the reference voltage though, and they will actually be amplified by the voltage gain of the operational amplifier. The stability of the reference voltage is therefore vital to the accuracy of a power supply of this type.

The Circuit

Figure 2.5 shows the full circuit diagram for the bench power supply unit. It is based on the L200 integrated circuit, which is a high quality device offering a variable output voltage and adjustable current limiting. It has comprehensive protection circuits, including output current limiting, thermal overload protection, safe operating area, and input voltage surge protection. It would be something of an exaggeration to claim that the device was indestructible, but it is certainly a hardy component that can take a certain amount of misuse without being damaged.

Figure 2.6 shows the internal arrangement of the L200. This is basically the same as the setup of Figure 2.4, and the "Error Amplifier" is the equivalent to the operational amplifier of Figure 2.4. The voltage reference is a high quality type which is fed from a constant current source in order to still further improve its performance. The "Pass Element" is the high power output stage, and it connects to the output via a current sense resistor. The higher the output current, the higher the voltage developed across this resistor. If this voltage

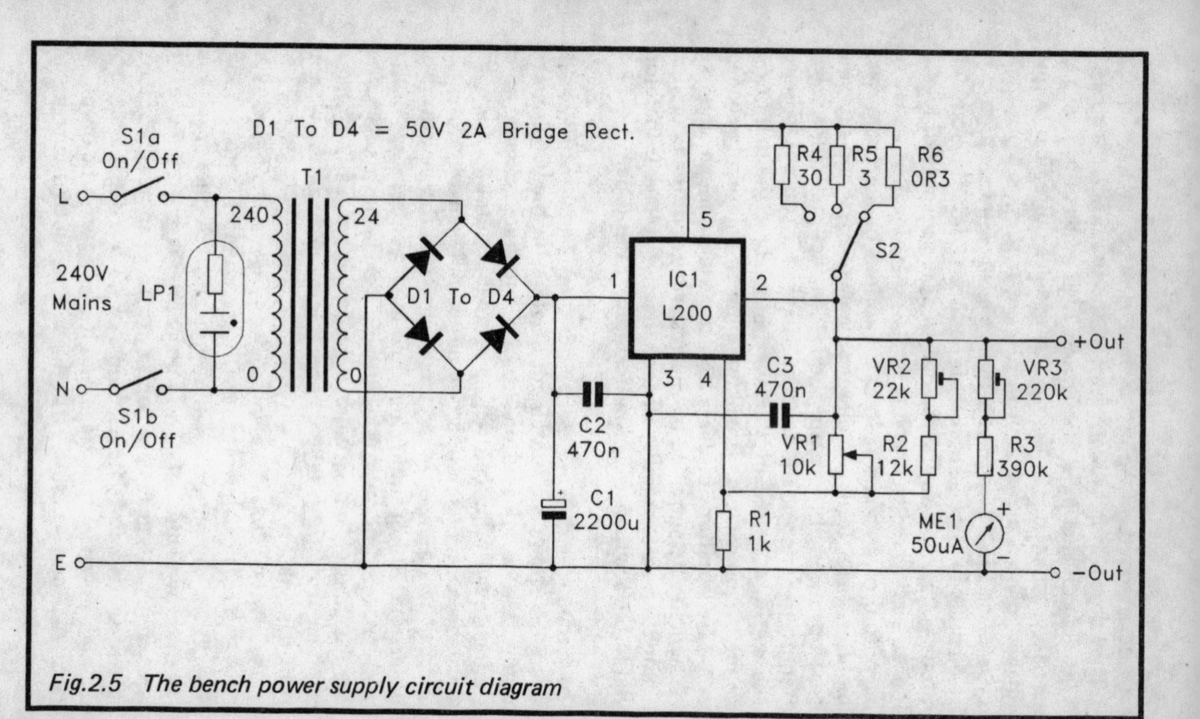

Fig.2.5 The bench power supply circuit diagram

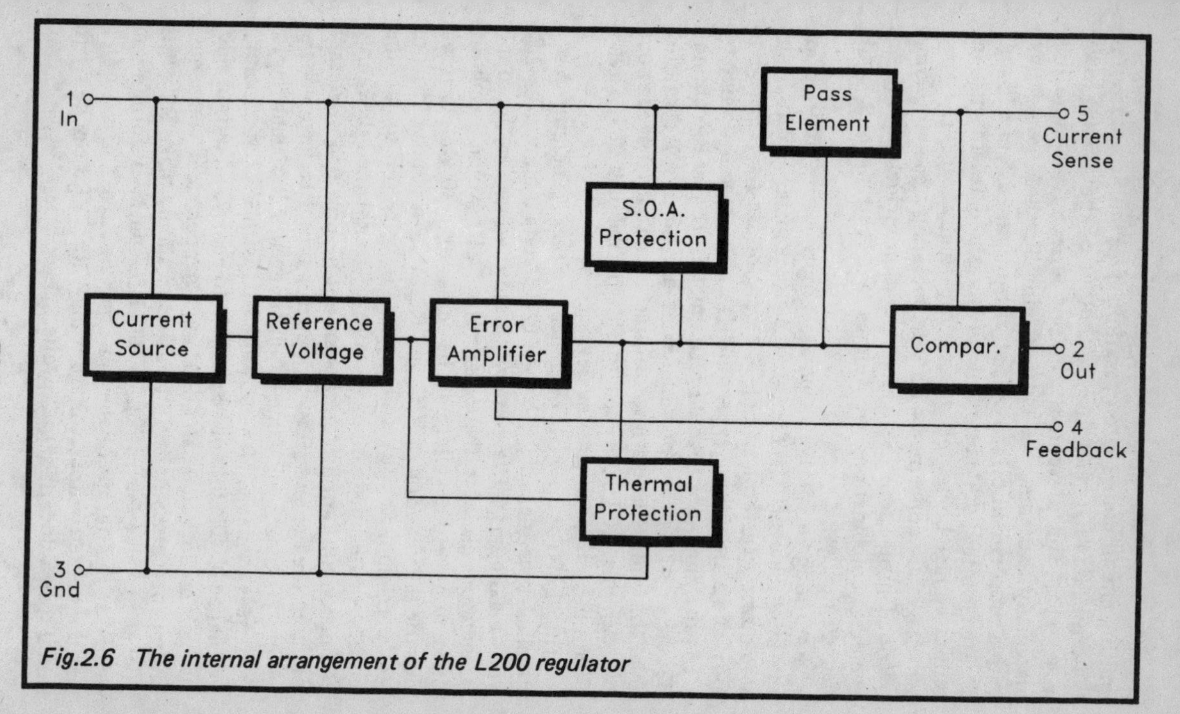

Fig.2.6 The internal arrangement of the L200 regulator

exceeds a certain threshold level (0.45 volts), the current limiting circuit comes into operation. This pulls the output of the error amplifier (and thus the output of the supply) lower in voltage. This prevents the output current from more than marginally exceeding the current limit threshold level. Even with a short circuit on the output, the output current will only slightly exceed the current limit threshold level, with the output voltage being pulled down to virtually zero. The other two stages are concerned with safe operating area and thermal overload protection. Safe area operation incidentally, means avoiding combinations of voltage and current that could damage the device.

Returning to the bench power supply circuit diagram, on the input side of IC1 there is the usual mains isolation and step-down transformer (T1), full wave bridge rectifier (D1 to D4), and a smoothing capacitor (C1). The specified value for C1 is the minimum I would recommend. If you can obtain a 3300μ or 4700μ component having a suitable voltage rating, then this should give a lower output "hum" level at high output currents. C2 is needed to aid the stability of IC1 (as is C3 on the output side of the device).

The L200's internal reference voltage generator has a nominal potential of 2.77 volts (2.65 volts minimum – 2.85 volts maximum). The minimum output voltage of the supply is therefore no more than 2.85 volts. VR1 would enable the maximum output voltage to be set at over 30 volts if it was not for the inclusion of VR2 and R2. VR2 is adjusted to limit the maximum output voltage to 20 volts, or some other desired figure up to about 25 volts. As explained previously, the maximum output current can not be provided at voltages of more than about 20 volts, but it is useful to have voltages in the 20 to 25 volt range available even if less than the full current is available. It is possible that with VR2 at maximum resistance it will still not be possible to achieve an output quite as high as 25 volts. If this should happen, and you require output potentials right up to 25 volts, simply make R2 a little higher in value (say 18k). VR3, R3, and ME1 form a voltmeter at the output of the unit, making it easy to accurately set the desired output voltage using VR1. VR3 can be set for a full scale value of 20 or 25 volts.

S2 is used to switch one of three current sensing resistors into circuit. R4, R5 and R6 give approximate limit currents of 15 milliamps, 150 milliamps, and 1.5 amps respectively. You may like to use different limit currents for the two lower limit levels, or to add more limit currents. The correct value for a given limit current is obtained by dividing 0.45 by the required limit current. The limit current should be in amps, and this gives an answer in ohms (or use milliamps to obtain an answer in kilohms).

Construction

Although the circuit of the bench power supply is quite simple, there are a number of awkward aspects when constructing the unit. Firstly, the unit is mains powered and the usual safety precautions should be observed.

IC1 has to dissipate quite a lot of power when the supply is operating with high output currents and low output voltages. It must therefore be mounted on a large heatsink. One having a rating of about 2 degrees Celsius per watt should suffice. What is probably a better alternative is to use a fairly large case of all metal construction to house the project, and to then use this as the heatsink. IC1 can be mounted direct onto a panel of the case or the chassis, but it will probably be more convenient to fit it on an "L" shaped bracket which is then bolted to the case at a convenient place. If you adopt this method the bracket must be quite large and made from a heavy gauge of aluminium (preferably 16 s.w.g.) so that it conducts heat into the case efficiently. There is no need to insulate the L200's heat-tab from the case since this tab connects internally to the "Ground" terminal, which will presumably connect to the earthed case anyway.

Several of the components require some explanation. With a maximum output current of 1.5 amps, the rating for T1 given in the components list (3 amps) might seem rather high. There is a problem with many variable voltage power supply designs in that the maximum permissible input voltage of the regulator is quite low, and the difference between the unloaded and loaded input voltages can be quite substantial. The maximum continuous input potential for the L200 is 40 volts. Using a 24 or 25 volt mains transformer the unloaded input voltage to the regulator will not be far short of this figure. In theory, multiplying the secondary voltage of the transformer by 1.41 gives the unloaded d.c. voltage, from which about 1 volt or so has to be deducted in order to take into account the voltage drops through the rectifier. In practice the actual voltage seems to be significantly higher than the figure derived from a calculation of this type. Multiplying by 1.5 or so and deducting nothing seems to give a more realistic figure.

The loaded d.c. voltage using a 24 or 25 volt transformer will be only about 24 or 25 volts. There are inevitably voltage drops through the series regulator circuit, and this permits a maximum output voltage with the full rated current at output voltages of about 20 volts and less. Using a mains transformer having a generous current rating helps to maximise performance at high output voltages. The rating is not as generous as it might at first appear. When using a bridge rectifier the current rating of the transformer needs to be substantially higher than the maximum d.c. output current that will be required. I would therefore recommend that T1 should not have a current rating of less than 3 amps. A higher rating is preferable, as the unit would then be able to supply the full 1.5 amp output current at output voltages of up to 25 volts. However, this would only marginally improve the usefulness of the unit, and higher current transformers would be very large, heavy, and quite expensive.

LP1 must be a panel neon of the type that has a built-in series resistor for operation on the 240 volt mains supply. R4 to R6 require values that are in the E24 series, but obtaining suitable components could be difficult. Not all component suppliers stock E24 series resistors of such low values. An easy solution to the problem is to make up each resistor from two or three resistors connected in series or parallel. As a couple of examples, R4 could consist of two 15 ohm resistors wired in series, or R6 could be made up from three 0.1 ohm resistors connected in series.

Ideally the output voltage would be monitored using a 20 or 25 volt panel meter. Panel meters of this type seem to be unavailable these days, making it necessary to use a current meter plus external series resistor in order to obtain a voltmeter action. The only real drawback of doing this is that the meter's calibration will be inappropriate. There is no real difficulty in converting the 0 to 50 calibration of the meter into corresponding output voltages, especially if the meter is set up to read 0 to 25 volts. However, the unit will be easier to use and will look neater if the scale of the meter is recalibrated. This is not too difficult, and with most modern panel meters the front unclips and then removing two small screws enables the scale plate to be detached. The existing lettering can be carefully scraped away using a tool having a sharp point, and then rub-on transfers can be used to add a 0 to 20 or 0 to 25 scale. Alternatively, the existing lettering can be left in place, with a new scale being marked on paper and then fitted in place over the original.

This is all quite straightforward in theory, but in practice you must remember that panel meters are very delicate mechanisms that are easily damaged. If you decide to alter the existing scale or fit a new one over it, exercise great care when working on the meter. If you are not reasonably skilled at working with delicate mechanisms you would probably be well advised to simply use the original scale.

Setting Up

Start with VR2 and VR3 at maximum resistance, and VR1 at minimum resistance. Using a multimeter to check the output voltage should produce a reading of around 2.65 to 2.85 volts. Adjust VR1 for an output of 20 volts, and then adjust VR3 for a reading of 20 volts on ME1. Next set VR2 at minimum resistance and VR1 at maximum resistance. Then set VR2 to produce the required maximum output voltage from the unit. The power supply is then ready for use.

Components for Bench Power Supply (Fig.2.5)

Resistors (0.25 watt 5% unless noted)

R1	1k
R2	12k
R3	390k 1%
R4	30R
R5	3R
R6	0.3R 1 watt

Potentiometers

VR1	10k lin
VR2	22k miniature preset
VR3	220k miniature preset

Capacitors

C1	2200μ 40V elect
C2	470n polyester
C3	470n polyester

Semiconductors

IC1	L200
D1–D4	50 volt 2A bridge rectifier

Miscellaneous

S1	Rotary mains switch
S2	3 way 4 pole rotary (only one pole used)
T1	Standard mains primary, 24 or 25 volt 3 amp secondary winding
ME1	50μA moving coil panel meter
LP1	Panel neon for 240 volt a.c. operation
	Large heatsink (see text)

Logic Pulser

A logic pulser is the digital equivalent to a signal generator for linear circuits. The analogy is quite a loose one though, since a signal generator is used to supply a signal to a chain of amplifiers or other processing stages, and the progress of the signal through these stages is then traced using an oscilloscope or a signal tracer. With digital circuits it is unusual to have a series of signal processing stages. Probably the only common example of a digital processing chain is a series of divider stages. The crystal calibrator described earlier uses a circuit of this type, and they are to be found in a few other applications such as baud rate generators for serial interfaces. With a circuit of this type a logic pulser can be used in a fashion that is similar to the way in which a signal generator is used with linear circuits.

Probably a more common use for a logic pulser is as a clock oscillator for a circuit that is under development, or when testing faulty circuits. It is often used to provide a very low frequency clock signal (sometimes less than 1Hz) so that the action of a circuit is slowed right down to make it more easily analysed. This method is not applicable to all types of logic circuit, and it will often not work properly with complex microprocessor based circuits. Where it can be used, it offers what is often a very good means of tracking down the defect in faulty equipment, or a design fault in a new design that is not performing as expected.

As a simple example of how a logic pulser might be used in practice, we will return to the divider chain example mentioned previously. One way of testing a circuit of this type is to disconnect the clock oscillator, inject the output from the logic pulser to the input of the divider chain, and then check the output signal from each divider stage using an oscilloscope. In practice this is an over involved method of testing the circuit, since the pulser is not really needed. You might just as well use the clock oscillator to provide the input signal to the divider chain!

Suppose that you do not have an oscilloscope though, but that you do have a logic probe. You can check with the logic probe that there is a pulsing output from each divider stage, and the probe might give you some idea of the mark-space ratio of the signal at each output. However, it will not give conclusive evidence that the signals are alright, and it may not show exactly what is wrong if a stage is faulty. Using a logic pulser to provide the input signal would enable a very low clock frequency to be used, and the output of the first divider stage would then operate at a rate which could easily be followed by looking at the "high" and "low" l.e.d.s on the logic probe. The outputs of subsequent divider stages might operate at too low a frequency, with long waits being necessary in order to check that the correct divider action is being obtained. To overcome this problem the output frequency of the pulser could be increased. As you check further along the divider chain the clock frequency can be stepped up in order to maintain a suitable output frequency from the stage being tested.

Many logic circuits contain more complex signals than those from simple divider circuits, but with everything slowed down to a suitable rate it is always going to be easier to check that suitable waveforms are present, and that the relative timing of signals are correct. Having two or three logic probes can be a definite advantage when undertaking this checking of comparative signal timing. It is also much easier to check the mark-space ratio of signals if they are at low frequencies. You can see from the on/off times of the logic probe's "high" and "low" l.e.d.s whether or not the mark-space ratio is approximately correct. This method is not as good as having an oscilloscope or a high quality logic analyser, but it is a good substitute if available funds will not accommodate the purchase of one of these units.

Circuit Operation

Ideally a logic pulser should cover a wide frequency range, with a large number of mark-space ratios available. Units of this type can be very complex, with the mark and space durations separately adjustable over wide limits and having crystal controlled accuracy. For most purposes this sort of sophistication is not required, and a more simple approach will suffice. This unit has the simple configuration shown in the block diagram of Figure 2.7.

The circuit is based on a little used form of oscillator known as a "ring" oscillator. Two negative edge triggered monostables are wired in series, and have the output of the second monostable coupled back to the input of the first one, completing the "ring" connection. Oscillation is achieved because as the output pulse from one monostable ends, it generates a negative edge that triggers the other monostable. The circuit therefore oscillates indefinitely, with first one monostable being triggered, and then the other.

Circuits of this type can be a bit unreliable, and in particular, they can sometimes be a bit reluctant to start. This problem is avoided here because when the supply is first connected, the trigger inputs of both monostables will be taken to outputs that will be at logic 0. This effectively provides a negative trigger signal that ensures the circuit starts to oscillate. It is possible that both monostables will trigger at switch-on. This will not matter, because the first output pulse to end will have no effect on the other monostable. It

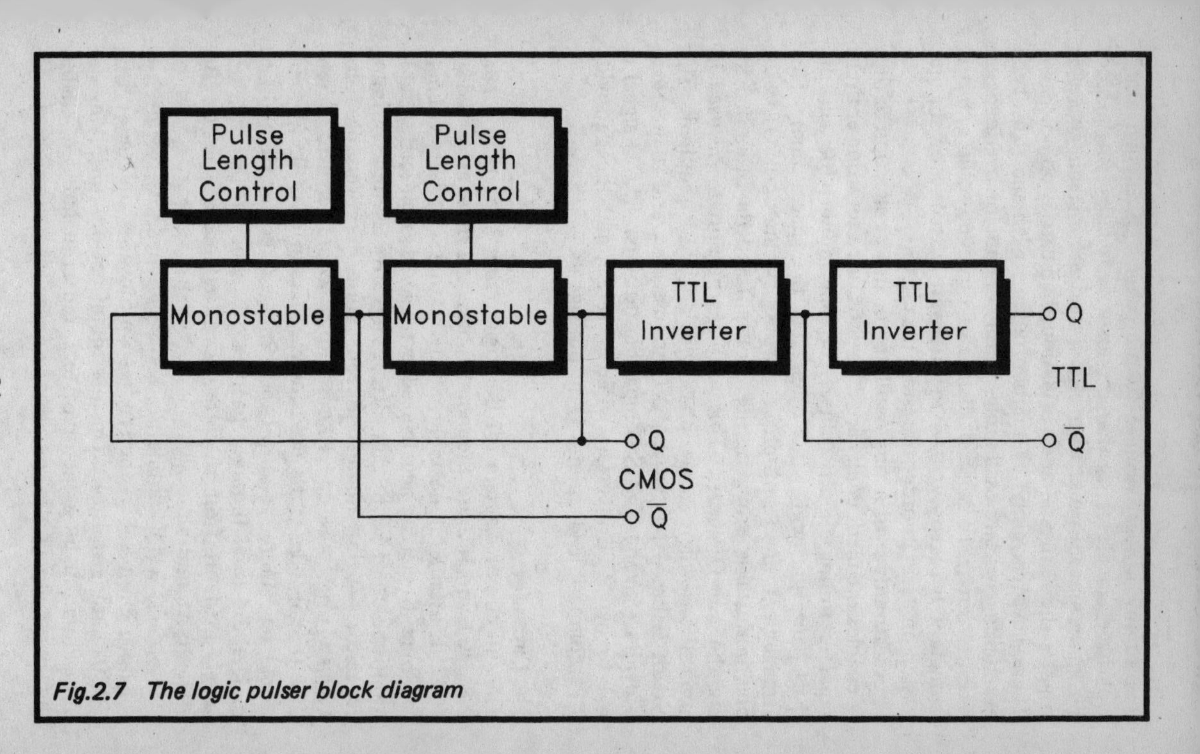

Fig.2.7 The logic pulser block diagram

will supply a negative edge to the other monostable, but it will have no effect as that monostable is already triggered. When the output pulse from the second monostable ceases, it triggers the first one, and oscillation then carries on normally. There have certainly been no problems with the prototype ever failing to start oscillating. Short circuits or overloads on outputs can result in an oscillator of this type stalling, but it is merely necessary to switch off, wait a second or so, and switch on again in order to restart the circuit.

An advantage of a ring oscillator in this application is that it enables the mark and space durations to be set using entirely separate timing circuits. In effect, one monostable sets the mark duration while the other sets the space time. This arrangement gives anti-phase outputs (Q and not Q), with one monostable providing the Q output and the other providing the not Q signal. This enables the unit to be used in applications where a two phase clock signal is required. The circuit is based on two CMOS monostables, and it therefore provides (5 volt) CMOS compatible outputs. The Q signal is fed to two TTL inverters wired in series, and these provide TTL compatible Q and not Q output signals.

The Circuit

The full circuit diagram for the logic pulser appears in Figure 2.8. IC1 and IC2 are the monostables, and these are based on 4047BE astable/monostable multivibrators which are connected here to act as negative edge triggered, non-retriggerable monostables. Each monostable has five switched timing capacitors, giving nominal output pulse durations of 10μs, 100μs, 1ms, 10ms, and 100ms. Logic would seem to dictate that for the 10μs pulse duration a capacitor of 100p in value would be required, and not the 22p specified for C1 and C6. In fact the total timing capacitance is about 100p, but it is largely made up from the internal capacitance of the 4047BE integrated circuit.

Normally a 43k timing resistor is switched into circuit, but for both monostables there is the option of switching over to a 430k timing resistor. This has the effect of multiplying the pulse duration by a factor of ten, and is included to enable the maximum pulse duration of each monostable to be

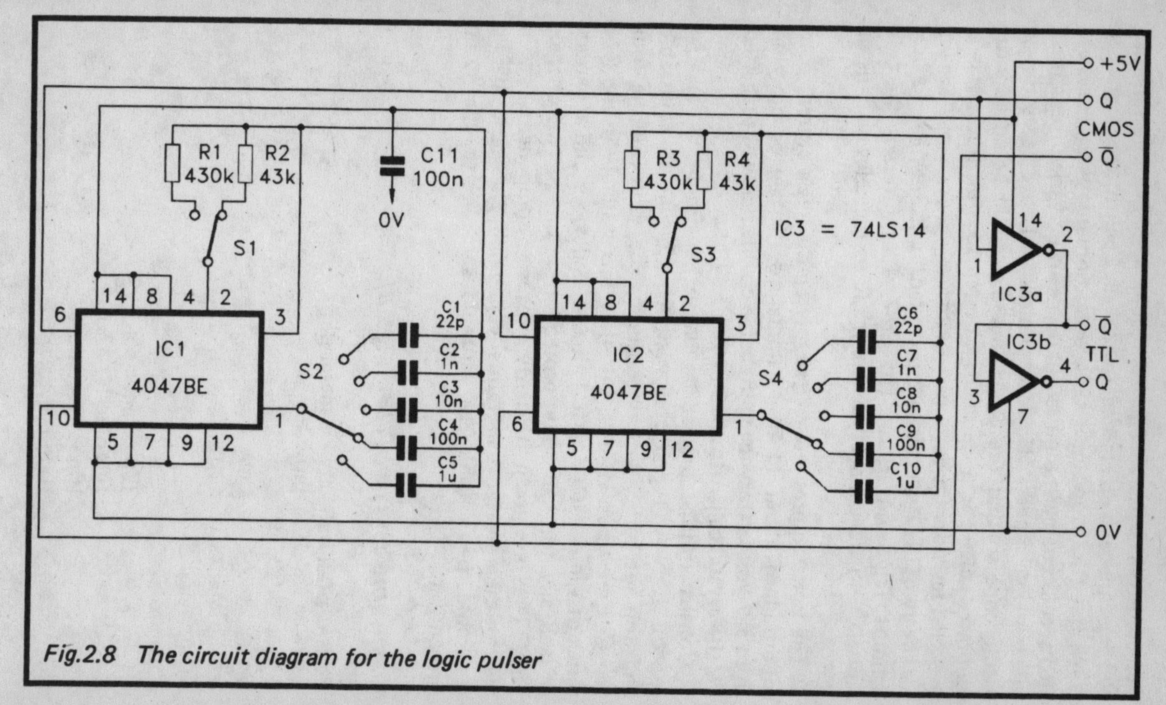

Fig.2.8 The circuit diagram for the logic pulser

increased to 1 second. The unit could be made more versatile by replacing each changeover switch and pair of resistors with a 39k fixed resistor and a 470k potentiometer wired in series. This would enable the pulse durations to be varied from 10μs to 1s in five ranges. However, unless you have access to an oscilloscope so that the potentiometers could be fitted with calibrated dials, it is probably better to opt for the two switched resistors for each monostable.

The two TTL inverters are provided by IC3, which is a hex Schmitt Trigger/inverter. The other four inverters are left unused, and no connections are made to the relevant pins of IC3.

The unit requires a 5 volt supply and it has a current consumption of about 11 to 12 milliamps. It can be powered from a 9 volt battery using the same circuit that was suggested for the crystal calibrator (see Figure 2.3).

Construction of the unit is reasonably simple. S2 and S4 can be 12 way 1 pole rotary switches having an adjustable end-stop set for 5 way operation. The values of the four resistors are unusual ones, but they are available in the E24 series of values. IC1 and IC2 are CMOS devices, and apparently lack any form of static protection on one or two of their pins. They are not particularly expensive components, but in view of their greater than normal vulnerability to static charges it would be sensible to scrupulously observe the usual precautions.

Components for Logic Pulser (Fig. 2.8)

Resistors (all 0.6 watt 1% metal film)

R1	430k
R2	43k
R3	430k
R4	43k

Capacitors

C1	22p ceramic plate 2%
C2	1n polyester 5%
C3	10n polyester 5%
C4	100n polyester 5%

C5	1μ polyester 5%
C6	22p ceramic plate 2%
C7	1n polyester 5%
C8	10n polyester 5%
C9	100n polyester 5%
C10	1μ polyester 5%
C11	100n ceramic

Semiconductors

IC1	4047BE
IC2	4047BE
IC3	74LS14

Miscellaneous

S1	s.p.d.t. miniature toggle
S2	5 way 1 pole rotary
S3	s.p.d.t. miniature toggle
S4	5 way 1 pole rotary
	14 pin d.i.l. i.c. holder (3 off)

Dynamic Transistor Tester

The transistor tester featured in Chapter 1 of this book, like most transistor testers, makes d.c. checks on the components under test. For most purposes this method of testing is perfectly satisfactory, but it is possible for a transistor to have good current gain but to still be less than fully operational. There is an interesting alternative form of transistor tester in the dynamic type. In other words, rather than making d.c. current gain checks, the device is tested using an a.c. input signal. This signal is usually at a frequency of several megahertz, where any deficiencies in the high frequency responses of test components will be revealed.

Figure 2.9 shows the arrangement used in this dynamic transistor checker. An r.f. oscillator generates the test signal, and this is followed by a buffer amplifier that ensures loading effects by the subsequent circuits do not cause the oscillator to stall. The device under test is placed in a simple amplifier circuit. It would be possible to switch the biasing, load resistors, etc. to suit n.p.n. or p.n.p. devices, but it is easier to simply have separate bias circuits and sockets for the two

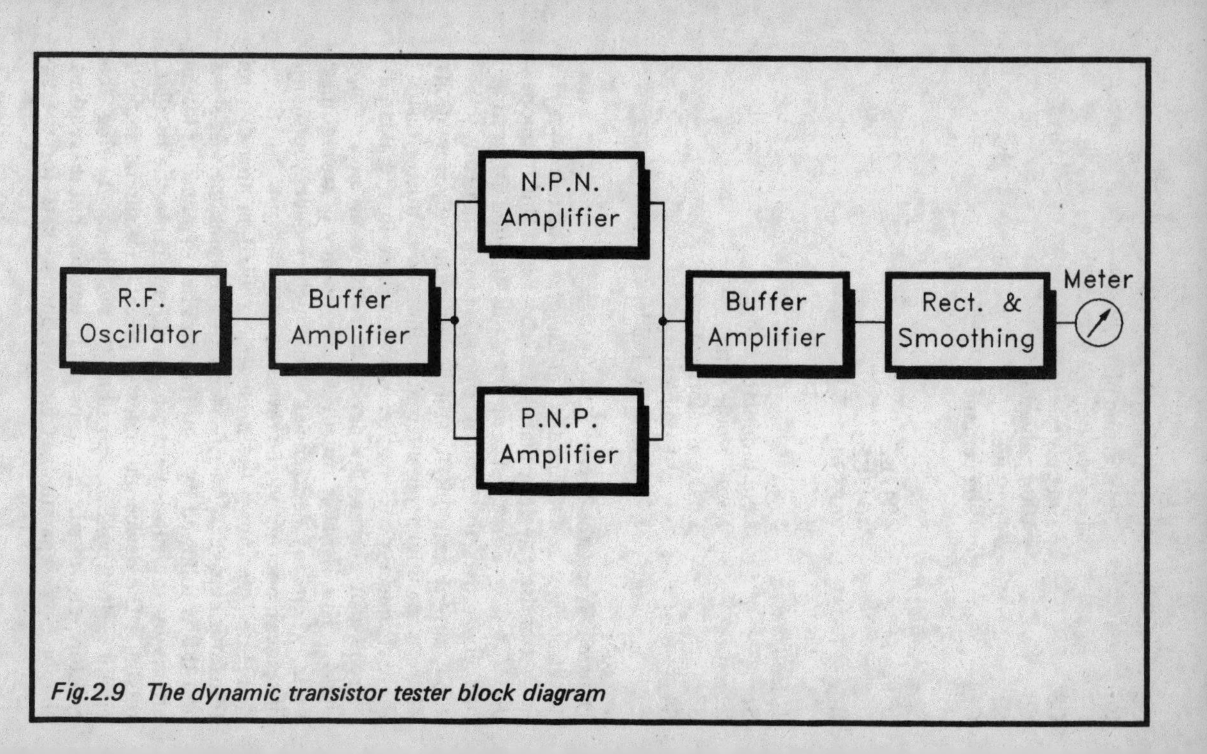

Fig.2.9 The dynamic transistor tester block diagram

types of transistor. The amplified signal is fed to a buffer amplifier which feeds a simple rectifier and smoothing circuit. This provides a d.c. output voltage that is roughly proportional to the a.c. output voltage from the amplifier, which is in turn proportional to the a.c. gain of the test transistor. A meter registers the d.c. output voltage, and the reading on the meter therefore gives an indication of how well (or otherwise) the test component is functioning.

There is a slight difficulty with a unit of this type in that it must provide correct biasing for any test device. There is no way of knowing what current gain the test components are likely to have, and enormous variations are quite likely. Low gain devices might offer a current gain of only about ten times, while the highest gain components offer around one hundred times as much current gain. It is important that all test devices operate at the same collector current. Otherwise there is a likelihood of some transistors seeming to offer excellent a.c. gain simply because they have a high d.c. current gain and are operating at high collector currents. The opposite would occur with low gain devices, which would operate at low collector currents and would consequently have low a.c. gains at the frequencies involved here.

One solution would be to have manually adjustable biasing so that the d.c. test conditions could be accurately set up at the correct levels for each component. This would be a bit slow and cumbersome though, and there are bias circuits which can largely negate the effects of different d.c. current gains. The bias circuit of Figure 2.10 was popular in the early days of transistor amplifiers. In those days the variations in current gain from one component to another were not the only problem. The germanium transistors in use at that time had what were often quite large leakage currents. For modern silicon devices the leakage level is usually just a fraction of a microamp. For germanium devices a leakage level of a milliamp or two would not be anything out of the ordinary. Variations in current gain plus wide variations in leakage level made it impossible to obtain accurate biasing using simple single resistor biasing, even if the type which utilizes negative feedback was used.

The bias circuit of Figure 2.10 guarantees accurate biasing

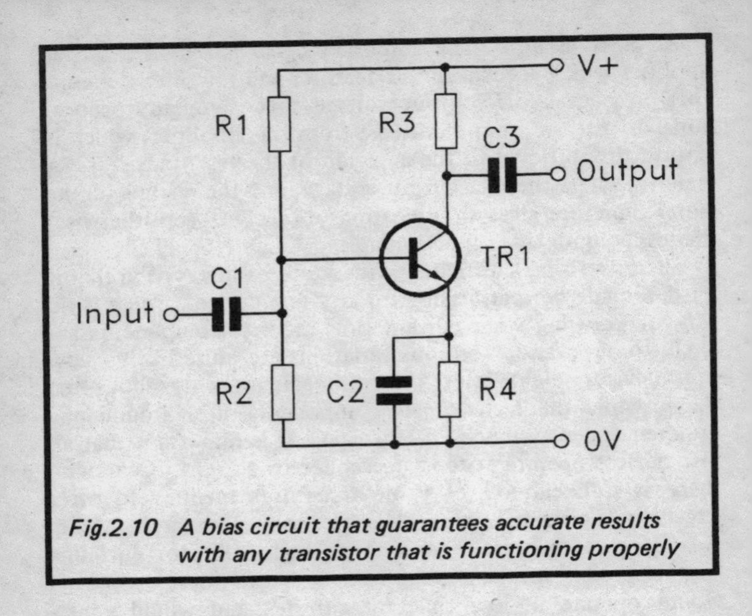

Fig.2.10 A bias circuit that guarantees accurate results with any transistor that is functioning properly

with any component that offers at least a moderate amount of current gain and which does not have a grossly excessive leakage level. Suppose that the supply potential is 9 volts, and that a bias voltage of about 4.5 volts is therefore needed at the collector of TR1. R3 could be made about four times higher in value than R4, and the biasing of the circuit would then be set so that there was about 1 volt across R4. As the collector and emitter currents of a transistor are not radically different, this would give about 4 volts across R3, which would leave about 4 volts across the collector–emitter terminals of TR1.

This gives more or less the required biasing. On the face of it the bias level at the collector of TR1 is a little high at 5 volts instead of 4.5 volts. However, the 1 volt across R4 reduces the maximum peak to peak output voltage swing of the circuit by 1 volt. Optimum biasing is therefore with the collector of TR1 half way between +1 volt and +9 volts, or at 5 volts above the 0 volt rail in other words. C2 bypasses R4 so that it does not introduce a.c. negative feedback that would reduce the a.c. gain of the amplifier.

For this method of biasing to be successful it is essential to be able to set the emitter voltage of TR1 at the correct level. This is not too difficult since the emitter voltage of a silicon transistor is about 0.6 volts less than base voltage. In our previous example where 1 volt was needed at the emitter of TR1, this would be achieved by using values for R1 and R2 that would bias the base of TR1 to 1.6 volts. The values of the bias resistors should be fairly low so that loading by TR1 does not significantly pull the base voltage away from the required figure. This is not a major problem since the emitter resistor introduces a lot of d.c. negative feedback which boosts the input resistance of the transistor to a relatively high figure. This method of biasing is therefore largely independent of the transistor's current gain. It is the values of the resistors which have a more major effect on the exact bias voltages.

The Circuit

The full circuit diagram for the dynamic transistor tester appears in Figure 2.11. The r.f. oscillator is a crystal type, and is much the same as the one used in the crystal calibrator unit described previosuly. I used a 12MHz crystal for X1, but any type having a frequency of about 10 to 13MHz should be satisfactory (but avoid overtone types, as in this circuit these will operate at their fundamental and not the appropriate overtone). Use of a crystal might seem a bit extravagant, but these components are available quite cheaply these days, and using an inductor would probably not provide a significant reduction in the cost of the unit. Like the oscillator in the crystal calibrator, the usual capacitors from the collector and base of TR1 to earth have been omitted. Stray circuit capacitances will almost certainly be sufficient to produce strong oscillation, but capacitors of about 6p8 can be added to the unit if necessary.

TR2 is the buffer stage, and is an emitter follower type. The output of the unit is likely to be slightly excessive, and so an attenuator is used as the emitter load resistor for TR2. This gives an output level that will produce a strong signal from any good quality test device, but clipping is unlikely to occur. C2 couples the output of TR2 to the amplifier in which the test device is connected. S1a connects the signal

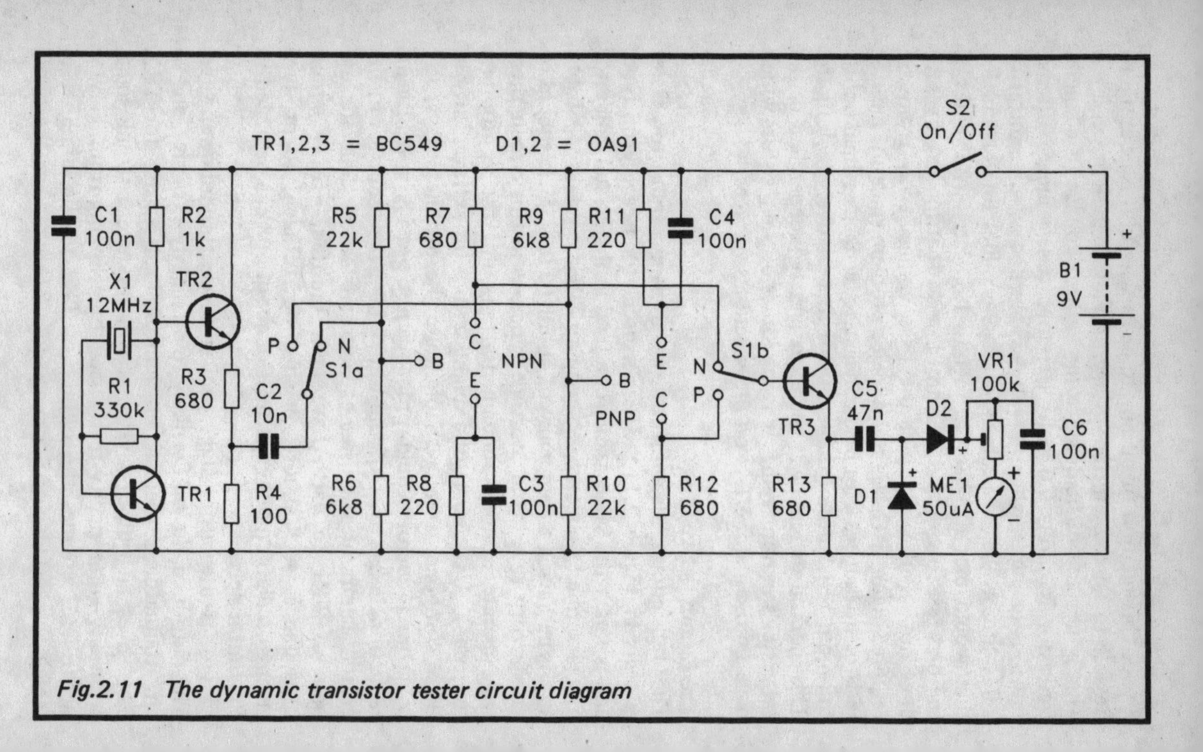

Fig.2.11 The dynamic transistor tester circuit diagram

through to the n.p.n. or the p.n.p. test amplifier, as required. S1b connects the output from the selected circuit through to the emitter follower buffer stage based on TR3. C5 couples the output from this stage to a conventional rectifier and smoothing circuit. This directly drives the meter circuit which has VR1 as the preset sensitivity control.

The component layout of the unit is not particularly critical, but avoid any long wires or obvious feedback routes which could adversely affect results. I used a good quality panel meter for ME1, but as readings are relative and the scaling is arbitrary, a low cost "level" or "tuning" meter should be perfectly satisfactory. These usually have lower sensitivities than the specified full scale current of 50µA, but it should still be possible to set the sensitivity of the unity at a suitable level by adjusting VR1. I would not recommend using a meter having a full scale current of more than 250 microamps, but most low cost meters seem to be 200µA or 250µA types.

The current consumption of the circuit is about 20 milliamps. If it is only likely to receive intermittent use a "high power" PP3 size battery is adequate as the power source. If the unit is likely to receive a great deal of use it would be better to opt for a higher capacity 9 volt battery, such as a PP9 type or six HP7 cells in a plastic holder.

The only adjustment needed to the completed unit is to set VR1 for a suitable sensitivity. This can only be given a final setting in the light of experience after the unit has been in use for a while. The optimum setting is one which does not allow any overloading of the meter on high gain components, but which gives something very close to full scale deflection of the meter when testing devices of this type.

Results using a unit of this type are quite interesting. Transistors of the same general type tend to give virtually identical readings, even though their d.c. current gains might be radically different. Devices which have high transition frequencies but relatively low d.c. current gains will usually give reasonably high readings when tested with this unit. At the frequency involved in this test it is more the gain/bandwidth product of test components that determine their level of performance, rather than their d.c. current gains. The unit will soon sort out any components which offer good d.c. current gains but which are faulty and offer poor a.c. gain.

Components for Dynamic Transistor Tester (Fig.2.11)

Resistors (all 0.25 watt 5%)

R1	330k
R2	1k
R3	680R
R4	100R
R5	22k
R6	6k8
R7	680R
R8	220R
R9	6k8
R10	22k
R11	220R
R12	680R
R13	680R

Potentiometer

VR1	100k miniature preset

Capacitors

C1	100n ceramic
C2	10n polyester
C3	100n ceramic
C4	100n ceramic
C5	47n polyester
C6	100n polyester

Semiconductors

TR1	BC549
TR2	BC549
TR3	BC549
D1	OA91
D2	OA91

Miscellaneous

S1	d.p.d.t. miniature toggle

S2	s.p.s.t. miniature toggle
ME1	50μA moving coil panel meter (see text)
X1	12MHz miniature wire-ended crystal (see text)
B1	9 volt (PP9 size – see text)

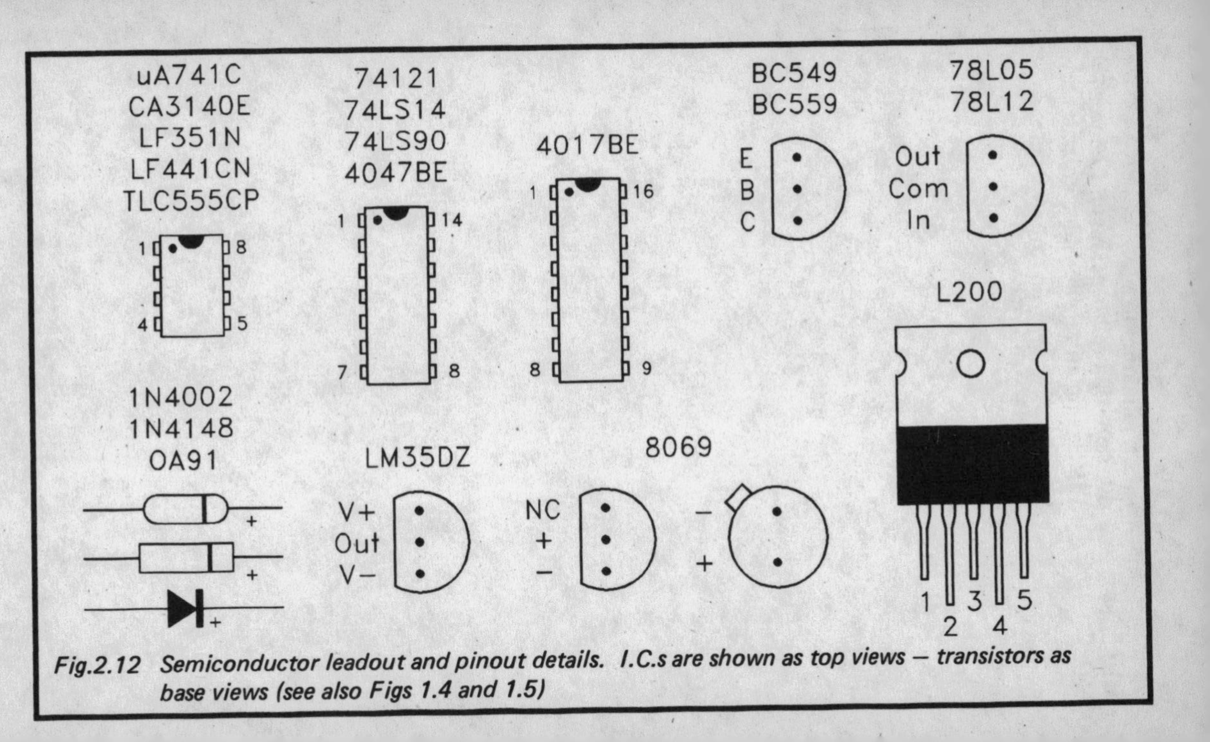

Fig.2.12 Semiconductor leadout and pinout details. I.C.s are shown as top views – transistors as base views (see also Figs 1.4 and 1.5)

Notes

Notes

Notes

Notes

Notes

Notes

Notes

Please note following is a list of other titles that are available in our range of Radio, Electronics and Computer Books.

These should be available from all good Booksellers, Radio Component Dealers and Mail Order Companies.

However, should you experience difficulty in obtaining any title in your area, then please write directly to the publisher enclosing payment to cover the cost of the book plus adequate postage.

If you would like a complete catalogue of our entire range of Radio, Electronics and Computer Books then please send a Stamped Addressed Envelope to:

BERNARD BABANI (publishing) LTD
THE GRAMPIANS
SHEPHERDS BUSH ROAD
LONDON W6 7NF
ENGLAND

160	Coil Design and Construction Manual	£2.50
205	Hi-Fi Loudspeaker Enclosures	£2.95
208	Practical Stereo & Quadrophony Handbook	£0.75
214	Audio Enthusiast's Handbook	£0.85
219	Solid State Novelty Projects	£0.85
220	Build Your Own Solid State Hi-Fi and Audio Accessories	£0.85
222	Solid State Short Wave Receivers for Beginners	£2.95
225	A Practical Introduction to Digital ICs	£2.50
226	How to Build Advanced Short Wave Receivers	£2.95
227	Beginners Guide to Building Electronic Projects	£1.95
228	Essential Theory for the Electronics Hobbyist	£2.50
BP2	Handbook of Radio, TV, Industrial and Transmitting Tube and Valve Equivalents	£0.60
BP6	Engineer's & Machinist's Reference Tables	£1.25
BP7	Radio & Electronic Colour Codes Data Chart	£0.95
BP27	Chart of Radio, Electronic, Semiconductor and Logic Symbols	£0.95
BP28	Resistor Selection Handbook	£0.60
BP29	Major Solid State Audio Hi-Fi Construction Projects	£0.85
BP33	Electronic Calculator Users Handbook	£1.50
BP36	50 Circuits Using Germanium Silicon and Zener Diodes	£1.50
BP37	50 Projects Using Relays, SCRs and TRIACs	£2.95
BP39	50 (FET) Field Effect Transistor Projects	£2.95
BP42	50 Simple LED Circuits	£1.95
BP44	IC 555 Projects	£2.95
BP45	Projects in Opto-Electronics	£1.95
BP48	Electronic Projects for Beginners	£1.95
BP49	Popular Electronic Projects	£2.50
BP53	Practical Electronics Calculations and Formulae	£3.95
BP54	Your Electronic Calculator & Your Money	£1.35
BP56	Electronic Security Devices	£2.50
BP58	50 Circuits Using 7400 Series IC's	£2.50
BP62	The Simple Electronic Circuit & Components (Elements of Electronics – Book 1)	£3.50
BP63	Alternating Current Theory (Elements of Electronics – Book 2)	£3.50
BP64	Semiconductor Technology (Elements of Electronics – Book 3)	£3.50
BP66	Beginners Guide to Microprocessors and Computing	£1.95
BP68	Choosing and Using Your Hi-Fi	£1.65
BP69	Electronic Games	£1.75
BP70	Transistor Radio Fault-finding Chart	£0.95
BP72	A Microprocessor Primer	£1.75
BP74	Electronic Music Projects	£2.50
BP76	Power Supply Projects	£2.50
BP77	Microprocessing Systems and Circuits (Elements of Electronics – Book 4)	£2.95
BP78	Practical Computer Experiments	£1.75
BP80	Popular Electronic Circuits - Book 1	£2.95
BP84	Digital IC Projects	£1.95
BP85	International Transistor Equivalents Guide	£3.50
BP86	An Introduction to BASIC Programming Techniques	£1.95
BP87	50 Simple LED Circuits – Book 2	£1.35
BP88	How to Use Op-Amps	£2.95
BP89	Communication (Elements of Electronics – Book 5)	£2.95
BP90	Audio Projects	£2.50
BP91	An Introduction to Radio DXing	£1.95
BP92	Electronics Simplified – Crystal Set Construction	£1.75
BP93	Electronic Timer Projects	£1.95
BP94	Electronic Projects for Cars and Boats	£1.95
BP95	Model Railway Projects	£1.95
BP97	IC Projects for Beginners	£1.95
BP98	Popular Electronic Circuits – Book 2	£2.25
BP99	Mini-matrix Board Projects	£2.50
BP101	How to Identify Unmarked ICs	£0.95
BP103	Multi-circuit Board Projects	£1.95
BP104	Electronic Science Projects	£2.95
BP105	Aerial Projects	£1.95
BP106	Modern Op-amp Projects	£1.95
BP107	30 Solderless Breadboard Projects – Book 1	£2.25
BP108	International Diode Equivalents Guide	£2.25
BP109	The Art of Programming the 1K ZX81	£1.95
BP110	How to Get Your Electronic Projects Working	£2.50
BP111	Audio (Elements of Electronics – Book 6)	£3.50
BP112	A Z-80 Workshop Manual	£3.50
BP113	30 Solderless Breadboard Projects – Book 2	£2.25
BP114	The Art of Programming the 16K ZX81	£2.50
BP115	The Pre-computer Book	£1.95
BP117	Practical Electronic Building Blocks – Book 1	£1.95
BP118	Practical Electronic Building Blocks – Book 2	£1.95
BP119	The Art of Programming the ZX Spectrum	£2.50
BP120	Audio Amplifier Fault-finding Chart	£0.95
BP121	How to Design and Make Your Own PCB's	£2.50
BP122	Audio Amplifier Construction	£2.25
BP123	A Practical Introduction to Microprocessors	£2.50
BP124	Easy Add-on Projects for Spectrum, ZX81 & Ace	£2.75
BP125	25 Simple Amateur Band Aerials	£1.95
BP126	BASIC & PASCAL in Parallel	£1.50
BP127	How to Design Electronic Projects	£2.25
BP128	20 Programs for the ZX Spectrum and 16K ZX81	£1.95
BP129	An Introduction to Programming the ORIC-1	£1.95
BP130	Micro Interfacing Circuits – Book 1	£2.25
BP131	Micro Interfacing Circuits – Book 2	£2.75